Praise for *Anatomy of a Fatal Fire*

AF808066

This deeply insightful book taught me much about being human and even more about normal reactions to abnormal incidents. But most importantly, it made abundantly clear that we firefighters are limited not only in our actions, but also in our thinking. This thought-provoking book is absolutely essential reading for every fire officer and each firefighter seeking a better understanding of how our mind works on the fireground. That knowledge is hugely important, because when disaster strikes, each one of us changes in a split second from a bunker gear superhero to a vulnerable human being with inherent mental flaws.

—Sgt. Edward Savat
Emergency Rescue Zone 1, West Flanders

An honest insight into the restless thoughts of a fire officer who has experienced the most painful verdict: the line-of-duty death of a colleague. Mental preparedness as a counterweight for sometimes overestimated physical preparedness. A must-read for every fire officer who hopefully never has to experience the hardest part of our job.

—Chris Addiers
Former chief fire officer, Antwerp Fire and Rescue Services;
former president of the Federation of European Fire Officers (FEU)

A brilliant book that meticulously explains what happened one fateful night in August. As you follow in the footsteps of our firefighters, your esteem only deepens. The most fascinating thing about this work is the journey through the mind on which the author takes you. You gain useful insights into how we think, and you will soon realize that in human decision making, countless assumptions and rationalizations—often made in a split second—make the difference between life and death. I have seldom read such a relatable and gripping book that presents facts as crystal clear and at the same time offers a deeply human explanation for our daily thoughts and actions.

—Luc Vrijdaghs
Managing director of the city of Beringen

A fascinating and mind-blowing book. If you only read one book about the fire service, it must be this.

—Raphael Verbinnen
Former warrant officer, Aarschot Fire and Rescue Services

We already know a lot about fire development, but this book emphasizes once more that human factors should not be underestimated by anyone working in the frontline. This is a personal story in which Marc dares to adopt a vulnerable attitude. It's impressive to read and to hear during his presentations at conferences.

—Bas Bor (fire officer) and Job Kramer (crisis manager)
Fire Service South Holland South, the Netherlands

As firefighters we make choices. Everyday. Every second. But we are all just human beings that function as human beings. We think, we have memories. Most of the time that helps us fight fires, but sometimes not. Marc describes an incident that is the nightmare of each firefighter. His experience and thoughts written down in this captivating book are something to keep in the back of your mind on every fireground.

—Hendrik Jellema
Officer, Fire Service Fryslân, the Netherlands

An impressive work that provides compelling insights into how the human mind works in normal circumstances and under difficult conditions. A must-read for not only firefighters and first responders, but for everyone as throughout everyday life our actions, thoughts, and feelings are more directed by the unconscious than you would think at first glance.

—Belgian National Fire and Intrusion Protection Association (ANPI)

Marc has experienced an incident that we all as incident commanders never hope to experience. Each and every person has to come to terms with such a tragic incident in their own way. The manner in which Marc tried to understand what has happened, and how he has managed to make sense out of it, is very inspiring. He turned the tragic incident into a learning experience. Well, how fortunate for us to have a colleague daring enough to share his compelling insights on the matter in an intense read.

—Tanja Guyken
Officer, Fire Service Twente, the Netherlands

If this particularly moving and compelling story teaches us one thing, it is for sure that not everything you see is what you think you see. Our perception is much more an own construction than a faithful reproduction of reality. How you make sense of situations is much more defined by what you believe, know, and understand than you might think at first sight.

—André van Halewyck
Chief editor, *Kritak*

In *Anatomy of a Fatal Fire*, Coenen aims to explain how things can go dramatically wrong in tackling fires and other emergencies, with profound lessons relevant to ordinary life. He draws from philosophers of science, neuroscientists, geneticists, and psychologists to show how we make decisions, and how these can be terribly mistaken. Using examples such as the Kaprun or Grenfell Tower fires and the murder of Amadou Diallo, he describes how our predictions can fatally mislead. Coenen modestly, but passionately, sets out a rational basis for a professional attitude to safety and responsibility. His candid views are poignant and instructive, not only for firefighters but also for anyone who gets it wrong from time to time.

—Geerdt Magiels
Biologist and philosopher of science

The level of self-reflection in this enthralling book is something you seldom see. It is not often that you find a fire officer who dares to ask aloud what he did wrong on one fateful night when two of his colleagues tragically lost their lives in a terrible blaze. This deeply insightful book is not only an impressive tribute to these fallen colleagues, it also contains an important lesson for firefighters and first responders all around the world: how we make sense of situations and how we make decisions is determined by context and intuition.

—Maarten van Severen
Head of the Network Fire Service, Flanders

A poignant story in all its facets. It is not only deeply human, but also a fascinating read from a psychological and neurological point of view. Should you ever experience a traumatic event, the insights in this book may help you make sense of the situation.

—Luk Jordens
CEO, Car & Truck Training, Hasselt

Anatomy of a Fatal Fire takes the reader on an emotional rollercoaster of a fire officer being confronted with his worst nightmare. The story of a seemingly controllable routine job escalating so badly puts the reader in the boots of the person in charge at that critical moment. What would you have done? It is a well-researched and beautifully written incident report with potentially life-saving lessons and experiences and is therefore a must-read for all firefighters!

—Rob Jastrzebski
Chief editor, *Een-Een-Twee Magazine*, the Netherlands

In his book *Anatomy of a Fatal Fire*, fire officer Marc Coenen wonders why one of his interventions could end so dramatically. The result is a deeply human document in which emotions alternate with in-depth research in an erudite way.

This incredibly well-written book offers a fascinating view of how our mind works in stressful conditions. Just as in daily life our actions, thoughts, and feelings are then more directed by the unconscious than we think, or even like. Here Coenen follows in the footsteps of star journalist Malcolm Gladwell, who showed in his book *Blink* that intuition is something very complex.

The author's traumatic experiences one August night and his subsequent obsessive or even neurotic search for answers ultimately prove to be a creative and purifying process, culminating in this wonderful book, that will fascinate a wide audience far beyond the fire service!

—Ludo Vanroy
Chief editor, *Fireforum Magazine*, Belgium

This book reads like a journey through the memories of a man with guilt. It is raw, honest, thoughtful, lucid, instructive. It is self-critical, it is a search for answers, and it is certainly also an attempt to move on with life. A staggering insight into head and heart.

—Rik van Puymbroeck
Journalist, *De Tijd*

Losing colleagues on a fireground: that is the fear of every firefighter. You return to the firehouse with more questions than when you set off, even when you have done nothing wrong, as is the case with the fire officer who wrote this book. Lucky for us, he had the courage to investigate thoroughly his performance that fateful night. His analysis goes beyond the classic technical approach you usually find in an incident analysis, as he focuses on our thought processes. Normally they help you tackle the fire, but they can just as well lead you astray. Truly a refreshing and enlightening approach to incidents.

A similar experience more than 30 years ago, which luckily ended less dramatically, made my read as intense as it was instructive. Each page brought back memories, but understanding the underlying processes helped to give everything its place.

This book, which reads like a novel and is flavored with touches of history, is a must-read for every fire officer who wants to understand how we think during interventions and to steer clear of the pitfalls inherent to the way our brains function. If you want to be better prepared when disaster strikes, this is the book to read!

—Eddy Goossens
Former fire officer, Geel Fire and Rescue Services

ANATOMY OF A FATAL FIRE

ANATOMY OF A FATAL FIRE

WHAT HAVE I DONE WRONG?

MARC COENEN

FIRE ENGINEERING
BOOKS

Disclaimer
The recommendations, advice, descriptions, and methods in this book are presented solely for educational purposes. Photos are for instructional purposes only. Always wear the proper level of approved PPE when conducting training drills and operating at incidents. The author and publisher assume no liability whatsoever for any loss or damage that results from the use of any of the material in this book. Use of the material in this book is solely at the risk of the user.

Copyright © 2024 by
Fire Engineering Books & Videos
110 S. Hartford Ave., Suite 200
Tulsa, Oklahoma 74120 USA

800.752.9764
+1.918.831.9421
info@fireengineeringbooks.com
www.FireEngineeringBooks.com

Executive Vice President: Eric Schlett
Vice President, Group Publishing: Amanda Champion
Acquisitions: David Rhodes, Diane Rothschild
Director, eLearning and Books: Starlet Franz
Production Manager: Tony Quinn
Book Designer: Robert Kern, TIPS Publishing Services, Carrboro, NC
Cover Designer: Trent Farar

Library of Congress Cataloging-in-Publication Data Available on Request

ISBN: 9781593705954

All rights reserved. No part of this book may be reproduced, stored in a retrieval system, or transcribed in any form or by any means, electronic or mechanical, including photocopying and recording, without the prior written permission of the publisher.

Printed in the United States of America

1 2 3 4 5 28 27 26 25 24

For all firefighters who,
wherever in the world,
have lost their lives
in the line of duty.

CONTENTS

FOREWORD

I was in Beringen at the time of the fateful fire. For me everything began in the early hours of that August night in 2019. In response to an emergency call, my colleagues and I rush out of the fire station of Lommel and head toward the city of Beringen. With our fire tanker truck we will assist in assuring the water supply for the fire trucks at the fireground.

On our way to Beringen, I am envisaging the scenario that awaits us. The operation is for sure going to take hours, and as usual we are not going to catch a glimpse of the fire itself. We will have plenty of time to make small talk with our colleagues of the other fire services, who are also heading toward Beringen in their fire tanker truck. All in all, a rather straightforward intervention.

Upon arrival, we deploy our resources. Then we are told that two firefighters are buried under the rubble of the collapsed roof of the burning supermarket, and a third colleague has been rushed to hospital with severe burns. Those messages strike home. The intervention is not at all going according to the script I had envisaged.

At the fireground I run into my colleague Marc Coenen. I realize that earlier in the night he must have been called to the scene as fire officer on duty. His state of mind is at an all-time low. He is struggling with a crushing feeling of guilt, and that inner battle is tearing him apart bit by bit. I try to give him some words of comfort, but he does not seem responsive anymore.

He is wandering around. Later that morning our ways part. He attends a debriefing, and I return to Lommel.

In the afternoon I give him a call. After several attempts I finally get hold of him. The conversation is downright awkward. Carefully I point out that he has hard times ahead. Coming to terms with such a traumatic event is by no means easy. But my words appear not to get through to him, like water off a duck's back. Clearly, he still has not yet fully grasped what has happened last night. The following day I call him again. His reaction is completely different. Now he talks my ear off, and I cannot get a word in.

The days after the fire I experience a Marc who is completely foreign to me. I call him almost every day, but each time our conversation is one-way traffic. He talks, and talks, and keeps on talking, and I do not get a chance to say a word. Day by day I am getting more and more worried about him. And rightly so: all the weird things he is trumpeting about no longer make sense to me. One of his strange ideas is even writing a book about the incident, as if that is what the world needs now!

And look here, I am impressed by the result. The first chapter reads like a script of a Hollywood movie. How incident commanders make decisions at the fireground is dealt with in the second chapter. With numerous examples, he argues convincingly that that process of decision-making can derail easily with all the potentially dreadful consequences it entails. In the third chapter, he links his psychotic experiences of the days following the fire to a dizzying trip through the history of mankind that takes us from the present, via a stopover in the Middle Ages, to the realm of the pharaohs and back. With here and there a touch of humor, he accurately portrays the myriad of pitfalls inherent to the intuitive mode of thinking that each one of us uses whenever in everyday life we make decisions under time pressure or stress.

And as for me, I find his wild idea of then not so strange anymore.

—Walter Cuypers
Officer of the Fire and Rescue Service
of North-Limburg (Belgium)

sedulo curavi, humanas actiones
non ridere, non lugere,
neque detestari, sed intelligere

With care I have tried
not to laugh at human actions, not to regret them,
nor to despise them, but to understand them.

—Benedictus de Spinoza

INTRODUCTION

My heart says one thing.
My head says another.
Very hard to get your heart and head together in life.

—Woody Allen

This book deals with the fatal fire that on Sunday night August 11, 2019, raged in an abandoned supermarket on Coal Mine Lane in Beringen-Mine (Belgium). A fire that destroyed several lives. That night, I was the fire officer on duty. The image I will never get out of my head is when I leave the fireground in the early hours of the morning: Two tents of the fire service are standing in front of a blackened wall of a building now reduced to ash and ruin. I know all too well that the bodies of two colleagues of mine are lying inside those red tents. They are dead. Earlier that night I had asked them to enter the building. They had to find out whether there were still any seats of fire in the building, and if so, they had to extinguish them. When I was being taken away from the fireground, I was already lost in my own head for several hours and wracking my brain over just one question: *What have I done wrong?*

In the months that followed, I began to read every book and article I could get my hands on. In an effort to come to terms with my overwhelming feeling of guilt, I turned to philosophers of science, historians, geneticists, molecular neurobiologists, behavioral biologists, psychologists, and a few winners of the Nobel Prize and the Ig Nobel Prize. At any cost I had to figure out how that operation could have ended so dramatically. You are holding the result of that research in all possible and impossible scientific disciplines in your hands. I have devoured numerous books on the brain and how it works, but the Word document "Oh happy day" on my computer is by far the most important source of this book. The document is named after a gospel song recorded by the Edwin Hawkins Singers in 1968. Like so many beautiful things in life, I stumbled upon it by accident. The impact of that particular song was so strong that the whole setting remains in my memory as vivid as it was the day it happened.

One Sunday morning I drive home from one of the football games of my sons Jonathan and Matthias. As always, I reverse into the driveway, for after having received an emergency call it is not only easier but also safer to drive out to the road. The car radio is tuned to Radio 1. Yes, I cannot help it. I am one of those persons looking for added value on the FM wave, and much to the annoyance of my wife and our sons, I am a die-hard listener of talk radio. One of the deals I have made with my sons is that when I have to take them to or from sports activities, I select the radio station, and that is always Radio 1, unless the added value is nowhere to be found, and strange but true, occasionally that happens.

Now back to my driveway. We are less than a month after the fire. For me, everything is far from back to normal. Where in the past reason was everything to me, now I am driven by emotions and even more emotions. I cannot make heads or tails of it. It seems as if the defense wall around my brain has been razed to the ground. Without the slightest resistance, everything can enter my wandering mind and touch me in the deepest part of my core. I am plagued with a feeling of guilt like I have never experienced before, and never hope to experience again.

Exactly at that moment the gospel song "Oh Happy Day" by the Edwin Hawkins Singers is playing on the car radio. It completely knocks me off my feet, as if the lyrics were written especially for me. Despite a brief career in the Catholic church as an altar boy, I hardly qualify as an ardent churchgoer nor am I particularly steadfast in the faith. But at that moment the lyrics struck me like a lightning bolt. Indeed, it must be a happy day when Jesus has washed your sins away.

Over time, the Word document "Oh happy day" grew into the biggest file on my computer. More than 112 pages covered the period from August 11, 2019, till August 14, 2020. As time went by, the number of entries not only decreased, but they also got shorter. At first glance, this Word document is nothing special. For sure I am not the first person, and I will definitely not be the last one, suffering from an acute stress disorder or another brain glitch for whom writing about those upsetting events not only improves his physical and mental health but along the way also develops in him a writing neurosis. Somewhere in the back of my mind, something else had to be at play. Or at least that is what my conscious mind tricks me into believing.

By training, I am a doctor in the ancient Near East, Egyptology. Yep, on a blue Monday I wrote a PhD at the Catholic University of Leuven. The subject of my doctorate was a particular kind of funerary text popular in Graeco-Roman Egypt. My first contact with a funerary papyrus of that type was in the early 1990s in the office of the professor who supervised my master's thesis. He had just returned from a study visit to Holland, and he came home with the new catalogue of the House of the Book in The Hague. Straight to the point he asked me if I knew Denon. I still vividly remember my first thought. At that time I was not yet the ardent listener of Radio 1 that I am now. Then my radio was tuned to a local radio station. With age comes wisdom. A supermarket had invested a lot of money in radio commercials. Time and again, a commercial was aired that ended with the words: "Quality foods Denon, in Brussels Street with parking lot in Wiering Street."

Unconsciously we take in more advertising than we are aware of. I almost recite the commercial word by word in response to the professor's question. But something in me—I suspect my intuition—told me that this for sure was not the answer he was hoping to get. At that time I could not assess his sense of humour, more specifically for my bland jokes, so I decided to keep it to myself and went for the safe answer: "No, I do not know him." A missed opportunity.

That day, the French diplomat, writer, artist, director of the Louvre, and above all, womanizer Dominique-Vivant Denon (1747–1825) entered my life together with an obscure type of funerary text, which would dominate my life for many years to come. The funerary papyri from Graeco-Roman Egypt have long since disappeared from my life, but Denon has stayed. At the age of 51, he managed, with a little help of Joséphine de Beauharnais, to secure a place in the *Commission des Sciences et des Arts à l'Armée de l'Orient*, which her husband Napoleon Bonaparte was about to send to Egypt in the wake of his expeditionary army. That day not only did this embedded war reporter *avant la lettre* come into my life but also an entire expeditionary force of roughly 54,000 men strong. Like Denon, countless soldiers and scientists could not resist writing about their adventures along the banks of the river Nile. As a result, we now have masses of information on every aspect of day-to-day life during the Napoleonic campaign in Egypt (1798–1801). But above all, those travel stories made me realize the tremendous value of eyewitness accounts.

There I stood mid-August 2019 with an all-consuming feeling of guilt and a brain in overdrive. I was in dire need of a scientific work that would help me sort out what was going on in my mind.

So I went to the psychologist with a thousand questions, but she was more interested in my answers to her questions than in answering all my questions on how the mind works. The two sheets of paper on which she rudimentarily explained that working were for me too valuable to let go to waste. Luckily I could take them home, and I still have them. However, her explanation did not fully still my ever-growing hunger for scientific information about how the brain operates under strain. According to her, everything I was experiencing fell within the range of normal reactions to an abnormal event. That was basically all she had to offer. Taking rest, that was what I needed the most according to her. But at that time, I was longing for a good book that would answer all those questions that were spinning round in my head at an ever-increasing dazzling speed. How broad is that range of normal reactions to an abnormal event? Where do those reactions come from? And what awaits me in the weeks and months to come?

Now, you are an Orientalist or you are not: "If the mountain won't go to Muhammad, then Muhammad must come to the mountain." If you cannot find a book that answers those questions, why as an experienced expert not write that book yourself? By documenting all the emotions that were rushing through my body and writing down all the things I was experiencing, I would have an excellent starting point. That was the idea. And that is how the Word document "Oh happy day" saw the light of day on my computer.

I was having more emotions and experiences than I actually could handle. And both were constantly reinforcing one another, in hindsight not so hard to understand. In the days and weeks after the fire, I experienced an unprecedented burst of creativity and impulsivity. A relentless flurry of ideas was triggered, insights were flooding my mind, and I wanted to rush into action. There was no stopping me. Everything, even the smallest things, seemed meaningful. No, they were meaningful. I was seeing connections everywhere. Coincidences ceased to exist. Everything happened for a reason and had to make sense. There had to be connections, if not causal or factual ones, then purely associative and irrational ones.

Other traits of my behavior during those days were socially unacceptable and emotionally clumsy. If I had something to say to someone, I picked up the phone and said what I had to say. Always unsolicited and brutally honest. Those conversations were more a rambling monologue on my part, as I simply could not stop talking and I constantly interrupted the person at the other end of the line.

Still, most of the conversations and encounters I had in the days after the fire were the most profound ones I have ever experienced in my entire life. Just like I discovered aspects of my personality that I was not the least bit aware of, so did I stumble upon traits of colleagues, friends, and loved ones that I never suspected them of having. It did not always work out right. Several times it came to a clash, even severe ones. When you take a dominant position and the other person quite rightly cannot agree with your umpteenth impulsive and short-sighted hunch, a clash is inevitable. However heated those arguments might have been at the time, in retrospect, I have learned more from them than from anything else.

In my search for explanations, they led me to the impact of elevated hormone levels on human behavior. Add to that already highly flammable cocktail an imbalance of several chemical messengers (neurotransmitters) inside your brain, and what do you get? Thoughts, feelings, and actions that I have never before experienced in my life, some of which are just too crazy for words.

This book is divided into three chapters. In the first chapter, I describe 24 hours in my life as a fire officer. I begin Saturday morning, August 10, 2019, and end the following day in the late morning. The turning point is Sunday night, August 11. Then at 3:08 a.m., from one moment to the next, a fire that seemed under control turned into an inferno. The impact was devastating. My life consists of two phases: a phase before and a phase after the fire. The phase before the fire was nicer, much nicer. That night my life changed and I will have to (learn to) live with it for the rest of my life.

In the second chapter, I try to figure out why that incident ended so dramatically and what was the trigger. Two firefighters lost their lives and a third one suffered severe burn injuries. In February 2020, the Federal Public Service Home Affairs concluded their research into the fire and released the incident report. One of the conclusions reads: "The different teams were deployed in a logical manner by the officer." How is it possible that despite my logical approach, the outcome was so terrible? In my opinion this points to a structural problem or a fundamental flaw in the way the fire service handles fires. I try to clarify that paradox in chapter 2.

The brain, the fire service, history in general, and that of ancient Egypt in particular run like a common thread through this work, but what really binds and holds it together are the conscious and unconscious modes of thinking and problem-solving in our predictive brain. We will look at those thought processes in the third and final chapter.

I am a doctor in Egyptology who now works as a fire officer. I can also formulate it in a negative manner. I do not hold a PhD in psychology, neurology, philosophy, or ethology, as behavioral biology is called nowadays. The only thing I as an Egyptologist have in common with a psychologist or neurologist is a few letters. More specifically, the last five letters: -*logist*. They go back to the Ancient Greek word *logos*, "reason, logic." And that can get you further than you might expect.

I have applied the scientific method that I have learned to master as a postdoctoral fellow in the ancient Near East to fields of research with which I have little affinity or none whatsoever. On top of that, I have read the wrong books. A scientist must always go back to the source. Instead of resorting to popular science books available in any self-respecting public library or that you nowadays with the greatest ease can order online to have delivered at your doorstep the next day, you have to work yourself a way through a pile of dry prose available in the shape of articles in internationally renowned academic journals or ground-breaking scientific works that most likely are written not in your mother tongue but in German, French, or English. I strongly recommend everyone to give it a go. As later will become apparent, I have a soft spot for (nor)adrenaline, but even love has its limits. What do you think of the following passage taken from a textbook on vitamins and hormones?[1]

> D. Phenylethanolamine-N-methyl transferase
>
> Noradrenaline in turn leaves the granules and is methylated to adrenaline in the cytoplasm. The methylation of noradrenaline takes place with S-adenosylmethionine as donor of the methyl group. This reaction is induced by glucocorticoids, which can reach relatively high concentrations thanks to a portal system within the adrenal glands. Adrenaline is then reabsorbed by another group of granules.

I have no idea what it means, but I take comfort in the knowledge that the grammar checking tool of my computer also cannot figure it out. In almost every sentence it proposes to modify at least two words. And underneath the title runs a long, red, wavy line. Still, the above text is crystal clear for persons trained in medicine, (bio)chemistry, or pharmaceutical sciences. Every field of research has its proper jargon. To fully master the jargon of the above-mentioned scientific disciplines is a complete study in itself that requires several years. I do not have that time, and I certainly do not want to spend the time that I still have on doing that. There are by far nicer ways to spend the time.

That is a problem you do not have with popular science books. You do have, however, another hurdle to overcome. Many of those books are merely popular and, from a scientific point of view, rather disappointing. I have tried to steer clear of unscientific positions by using various books written by different scholars, each holding an impressive scientific track record. That was at least the impression I gained from the myriad of academic titles and positions displayed on the back cover of their books. In doing so, I have probably managed to avoid one or two pitfalls, but several others undoubtedly remained under the radar. With this book I do not pretend to be the specialist in the working of the mind under severe stress. I just wanted to find explanations for the various weird things I was experiencing in recent scientific insights into how the brain works. That turned out to work quite well, but even more important to me was the realization that I was not going mad. All my physical reactions and racing thoughts turned out to be normal reactions to an abnormal incident.

Accidents are unfortunately common. They will continue to happen, both within and outside the fire service. If my story can offer support to a colleague of a fire department or another emergency service after a traumatic experience, then I have achieved my goal. This book is for sure not a self-help book for an acute stress disorder, a posttraumatic stress disorder, or any other stress-related condition. It can never replace the help that professional psychologists or psychiatrists provide. Should this book ever have had any therapeutic effect, then it has been completely used up by me. Writing this book allowed me to reconstruct, interpret, and integrate into my life the facts, observations, and feelings related to the most traumatic experience I endured in my professional career. This work is even less of a scientifically sound study on how the brain works under strain or on the process of decision-making. By writing down my experiences, perceptions, and ideas, I just offer food for thought to anyone who wants to read it. Whether you are interested in the working of the human mind, decision theory, or history, to me that does not matter. Anybody and everybody is welcome to this search for answers.

1

TWENTY-FOUR HOURS IN THE LIFE OF A FIRE OFFICER

1. SATURDAY, AUGUST 10, 2019, 11:00 A.M.

My eldest son Jonathan and I are off to a football tournament in the city of Tongeren. It is his first tournament with his new team. Joining that team was a good move, because in the late afternoon we bring home the cup. What a promising start to the new season! Back home I go to the village of Beverlo. There all weekend long Mid Harvest is celebrated with numerous popular festivities. I have to check whether the two tents they have put up comply with fire safety regulations.

After the check is completed, my wife, Grace, and Jonathan come to Beverlo. With all the food trucks and food stalls in the village center, we will for sure find something to our liking. Our youngest son, Matthias, could not join us. Grace has already dropped him off at the slumber party of his school friend. Tonight, they are going to sleep outside in a tent. For them the night cannot start soon enough.

There is a pleasant hustle and bustle in the village center of Beverlo. After having eaten more junk food than is good for me, I go to the sports hall nearby. Like every year, cycle ball club The Black Gold (*Het Zwarte Goud*) has asked me to wait tables. In addition to football, both our sons also play this totally unknown sport that combines two of the most popular sports in the world, for cycle ball is football being played on purpose-built bicycles.

There is never a shortage of sports activities at our home during the weekends. This year the club has booked a band to play in a trailer that stands on the terrace of the sports hall. In my grey T-shirt I stand out among the other crew members in their red club shirts. My club shirt lies in the closet at home. My pager is also at home. My current status in the paging system is "not available." During Mid Harvest it is always so busy in Beverlo that taking part in interventions of the fire service is simply out of the question. For the president of the club my grey T-shirt is fine. Grace, however, proposes to go home and fetch my club shirt.

With two sons playing cycle ball, we have got more than one team shirt at home. As Murphy's law dictates, my wife returns with a shirt in the wrong size: female medium. That is not really my size, but still I go and change in the men's room. My torso is not the most broad-shouldered, but a female medium is very tight. The shirt hides absolutely nothing, and that is no longer pleasing to the eye. I change back into my grey T-shirt and return to the cafeteria. Again Grace proposes to go and get the shirt but this time in the right size. Before she leaves, I ask her to also bring along my pager. Little did I know that such a trivial question would change the trajectory of my life.

It is all but crowded in the cafeteria of the sports hall. As usual the president has asked more members to come and help than necessary, so I am just wasting my time sitting at the bar. There is no work to be done. In that case I can better make myself useful for the fire service, and should I have to respond to an emergency call, then I can leave the cafeteria and go home after the intervention. No sooner said than done, my wife brings my pager to the sports hall together with the club shirt in the right size. Again I go to the men's room to change, and with the smartphone I alter my status in the paging system to "available fast." When a little later Grace decides to return home, I give her my smartphone. The battery is dead. From then on I no longer can modify my status in the paging system.

Around 10.00 p.m. the terrace of the sports hall fills up quickly. The band plays too loudly. The music is not to my liking, but finally there is work to be done. Sometimes there is even too much work. I have my hands full with clearing tables and occasionally I take an order.

2. SUNDAY, AUGUST 11, 2019, 2:00 A.M.

Around 2:00 a.m. the terrace and the cafeteria of the sports hall are as good as cleaned up. I ask the president if I may go home."No problem. Come back tomorrow," he says, "but then you won't have to stay this late, because that band is not going to play that long." On my way home I drive past the supermarket that has been languishing for years in the center of Beringen-Mine. It is 2:08 a.m. and I notice nothing out of the ordinary. I keep my eyes on the road and focus on the little traffic ahead. I pass the traffic light at the intersection of *Kasteletsingel* and *Nijverheidsweg*. But before I have reached the traffic light on the bridge over the Albert Canal, my pager begins to beep. The light is red. I get into the left-hand lane to turn left and stop. The light changes, and I turn around and rush to the fire station nearby.

To stop the beeping in my trouser pocket, I turn off the pager. I did not read the message "2:11 Ind (C10) (Off): BS Fire in the building next to the newspaper shop of Laenen, 2nd floor flames." Although I am in no time at the fire station, some colleagues have already arrived before me. I ask them where the fire is and they answer: "Coal Mine Lane." I think to myself: "That is strange. I just drove down that lane. Nowhere did I see flames or smoke, let alone a burning building. This is not going to be a big deal."

I get into my firefighting suit, jump in the command vehicle, and head toward Coal Mine Lane, house number 273. In the dark, house numbers are not easy to spot, but then I notice at the roundabout near the town hall two police vans with their blue flashing lights on. They must be responding to the same emergency call. Now I just have to follow them. Upon arrival, the police vans block the road in front of the supermarket. I drive past the building and park the command vehicle in the left lane next to one of the police vans. It is 2:18 a.m. I inform the dispatcher that I have arrived at the fireground.

Coal Mine Lane 273: I am familiar with this building. I have been here before. This time a blazing fire is raging in the empty store (building 3, fig. 1) between the former flower shop (building 4) and the family home (building 2) that, like the burning building, were once part of the old Delhaize supermarket of Beringen-Mine.

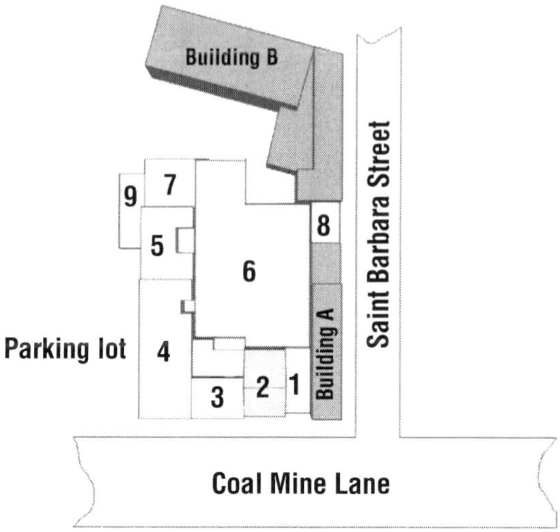

Figure 1. The fireground: The old Delhaize supermarket and its surroundings
Note. Floor plan by Toon Dumoulin

3. FROM WELL-BUILT TO SHODDILY RENOVATED

Like countless retail estates in Belgium, this supermarket, which once housed the Nopri and later the Delhaize, is the result of a complex organic growth spread over a long period of time. As such it comprises a haphazard mishmash of parts of old family homes and new commercial and residential blocks, with overlaid alterations and adaptations that had been made over the past few decades. On the ground floor more and more walls between the different buildings were knocked out or opened up as time went by. This resulted in a store of roughly 2,500 m^2 (50 m × 50 m) that had its heyday between 1960 and 2000. Several times plans were drawn up to give the exterior walls a complete makeover and to provide the supermarket a more appropriate look and feel. Each new plan was more ambitious than the last. The only thing all these building plans have in common is that they were never carried out. From the outside, nothing suggested that over time the ground floor had been converted into one large space. Not only did all fire compartmentation vanish; the fire resistance of the structure as a whole was also downgraded significantly.

The numbering of the various buildings that I use throughout this work, is outlined in the floor plan in figure 1 and table 1. The designation front refers to the facade on Coal Mine Lane. To the left of the building is a spacious parking lot, hence the name parking lot for the left side. Backside speaks for itself, and for the right side I use the name of the street where that part is situated: Saint Barbara Street.

In table 1, *C* stands for Coal Mine Lane or front, *P* for parking lot or left side, *B* for backside, and *S* for Saint Barbara Street or right side. The ground floor of buildings 1 to 9 form together one large supermarket. Buildings A and B are not parts of this retail estate but adjacent buildings. From Saint Barbara Street a narrow passage leads to the back of the supermarket. A shift in its axis, however, gives the impression that the narrow passage is nothing more than a small corridor that ends in a blind wall.

The oldest parts of the supermarket (buildings 1 and 2) date to the early 1950s. The type of plasterboard found on site after the fire suggests that the

Table 1. The buildings of and around the Delhaize supermarket

Building number	Description	Side	Year of construction	Measurements
Building A	Newspaper shop	Corner S/C	2003	
Building 1	Storeroom of a house knocked down in 2003	C	1950s	5.1 m × 12.5 m
Building 2	Family house with saddle roof	C	1950s	6.5 m × 12.5 m
Building 3	3-story commercial block with flats on the first and second floor	C	Late 1950s–early 1960s	10 m × 14 m
Building 4	3-story commercial block	Corner C/P	Late 1950s–early 1960s	30 m × 10 m
Building 5	3-story residential block	P	1960s	12 m × 10 m
Building 6	Store expansion		1960s	23 m × 36 m
Building 7	Service entrance of supermarket	B	1960s	9 m × 19.5 m
Building 8	Technical area (cold storage)	S	1960s	5 m × 8 m
Building 9	Storeroom in sandwich panels	Corner P/B	2007	4.4 m × 17.3 m
Building B	3-story residential block	B	2018	

shop extension (building 6) was constructed in the beginning of the 1960s. In its heyday, the supermarket incorporated buildings 1, 2, 3, 6, 7, and 8. In 2007 the shop was extended and renovated one last time. Only then buildings 4 and 5 ceased to be separate building volumes. Their ground floor was drawn into the retail space of the Delhaize. And partly in front of the residential block (building 5) and the service entrance (building 7), a storeroom in sandwich panels (building 9) was constructed on the side of the parking lot.

Typical of the renovations carried out in 2007 is the appalling lack of professionalism. The load-bearing wall on the inside of building 4 was opened up. One of the pillars put together by the construction workers speaks for itself. The fire consumed the fiberboard casing around the pillar, revealing its interior. The pillar looks like it is ready to be filled with concrete, but that concrete was never poured. Steel beams way too small to withstand the pressure of the upper floors are held in place by a simple strut, some loose nets of reinforcing steel, and plenty of air. They also worked on the flat roof. Several layers of thick insulation boards were placed on top of the original tar and gravel roof, which in their turn were covered by a new layer of tar and gravel. All the skylights and ventilation holes of the old roof were covered up. The night of the fire, the fire service would pay dearly for the work poorly executed in 2007.

4. PREVIOUS FIRES IN THE BUILDING

I was inside this building in the late afternoon of January 10, 2018. Already for several years the completely boarded up store just stood there languishing away in the center of Beringen-Mine. When a pedestrian walks alongside it, he notices the smell of gas and notifies the gas company. Their technicians wish to inspect the gas connection in the basement, but they cannot enter the building.

After consultation with the police, the fire service is called to the scene and we are asked to provide the technicians access to the building. The boarded-up entrance to the shop on the side of the parking lot (building

5) is made free. Behind the fiberboards, a double sliding door emerges. The search for the gas meter leads us to the basement, but we cannot enter it. It is awash with something that passes for water. In the end, we succeed in locating the gas leak at the front of the building. There appears to be a leak in the gas pipe under the pavement. Over the next few days the gas company will remove the underground connection. We board up the shop entrance and return to the fire station.

From the autumn of 2018 onward, the condition of the building deteriorates markedly and it is impossible to see past it. One fire follows after another and they get worse and worse. The situation gets out of hand more and more. The derelict supermarket in the center of Beringen-Mine slowly but surely turns into an eyesore.

On November 4, 2018, a squatter makes a fire in an iron barrel in one of the dilapidated flats on the first floor of building 5. He keeps the windows shut and the door is barricaded with a pallet. When firefighters enter the smoke-filled apartment, their carbon monoxide detector immediately goes into alarm. The squatter is prepared to leave the flat, but he strongly objects to being taken to hospital. We carry the barrel out of the building and extinguish the fire.

On November 20, 2018, we have to deal with a blazing fire in the leftmost flat on the first floor of, again, building 5. Coincidence or not, at that moment I am also driving alongside the run-down supermarket, but this time I am heading in the opposite direction, toward the village of Beverlo. As I approach the building, I see black smoke floating over Coal Mine Lane. There is something wrong. I park my car alongside the road and walk into the parking lot of the shopping complex. Flames are already shooting out of the flat's window. A bystander claims that there is still a person inside and that another one has entered the building to go fetch him. I dial the emergency number 1-1-2 and wait for my colleagues. Upon arrival, I share all the information I have gathered in the meantime, whereupon I continue on my way to Beverlo. Nobody was found inside the building.

During Christmas night 2018, again a fire erupts. This time pitch black smoke is coming from under the roof of building 2, the family home from the 1950s situated at the front. Shortly after the arrival of the fire service, the situation escalates into a spreading roof fire. The masses of cash register rolls and other papers carelessly left in ring binders or cupboards in the attic take their toll. After the fire, the front side of the gable roof is completely burned away and the rear side severely damaged.

5. SUNDAY, AUGUST 11, 2019, 2:18 A.M.

Now back to the last and fatal fire in the dilapidated supermarket. I arrive at the fireground at 2:18 a.m. The blazing fire at the front immediately catches my eye. You simply cannot miss it. What a difference from my passage just a few minutes ago! All in all, the flame front is of a reasonable size, and fortunately the seat of the fire is not situated deep inside the building but more at the front site. Before the arrival of the first fire truck, I take a quick look down Saint Barbara Street. Further down the road, I see smoke quietly flowing from some windows high in the wall of building 8. The speed of the smoke plume fits the picture. The blazing fire at the front not only produces a lot of smoke but also puts the entire building in overpressure. Because of this difference in air pressure, all that smoke is gently being pushed out. My first assessment is arson, because gas and electricity have been cut off for years. There might possibly be several seats of fire in the building, which is not uncommon with arson.

The situation can best be compared with the now torn-down Freinet school in the city center of Beringen. In that building the fire service also had to intervene repeatedly. Each time it was arson, and on several occasions it was alleged that one or more squatters might still be inside the building. Also this time the police tell me that there is a squatter in the building. I take note of it, but their information does not affect how we are going to fight the fire. To maximize the chances of survival of a person inside a burning building, the baseline of the fire service is now "put the fire out first." By extinguishing the fire as quickly as possible, we improve not only the chances of survival for the person inside the building but also our own working conditions. And should we come across any victims during extinguishing, then of course we will evacuate them immediately.

Shortly after my arrival at the scene, I can see and hear the first fire truck approaching from afar. I wait for my colleagues at the front of the building. At 2:19 a.m. the fire truck of Beringen is at the fireground. The ladder truck arrives shortly after. I instruct the warrant officer of the fire truck and inform him that there might be a squatter inside. He will have to deal with the fire at the front of the building. The ladder truck is instructed to use the monitor for extinguishing the fire on the first floor. Then I conduct my 360-degree initial size-up around the building. Across the parking lot I go to the back of the building and back again across the parking lot to the front side and into Saint Barbara Street. I will walk around the building in this way over and over again. I am not aware of the narrow passage at the back of the supermarket that gives access to Saint Barbara Street.

That night, the various facades of building look as follows:

- Front side on Coal Mine Lane with buildings 1, 2, 3, and 4: I can only look inside the building at the front of the supermarket. The shop windows of this side are still partly covered up with white metal boards. Some of these boards have already been pulled down so that we can begin extinguishing the fire. Behind the seat of the fire, the visibility inside the building is good, even very good. The seat of the fire is located at the front on the ground floor and the first floor. The fire is restricted to the back left corner of building 3. It is a raging fire.
- Left side or parking lot with buildings 4, 5, and 9: I can only take a look inside through the first shop window of building 4. No flames are visible. The fire is raging on the other side of the interior wall. All the other shop windows are either covered up with white metal strips, boarded up with fiberboard, or obscured with black foil stuck onto the inside of the window (buildings 4 and 5). The storeroom erected in sandwich panels (building 9) at the rear end of the parking lot is fully closed. It has no windows and the door is locked.
- Backside with buildings 6, 7, and 9: The back of building 9 has a white sectional garage door. The sandwich panels of building 9 join the wall of building 7, which is erected in concrete blocks. Roughly in the middle, that wall makes a right angle inward. In that corner there is a

narrow opening that is sealed off with a wooden board. From the rear I have a restricted view of the flat roof of the store extension (building 6).

- Right side on Saint Barbara Street with building 8: This side consists mainly of the elongated, red brick wall of the newspaper shop (building A). Next to it is a technical area of the Delhaize that once housed the cold storage (building 8). This outer wall in red bricks has a door and a garage door on the ground floor and three long horizontal windows on the upper story. To its right stand some of the flats that are part of the residential block (building B) constructed by the Cantonal Building Company behind the supermarket.

Grey-black smoke constantly pours out of various openings. On each side I have one reference point for reading the fire. I will discuss this technique used in firefighting later in this book. For now it suffices to say that this technique allows a fireground commander to assess the development phase of a fire and predict its further progress. Where exactly does that grey-black smoke leave the building?

- Most of the smoke comes out at the front. Perfectly logical, because the seat of the fire is at the front of building 3.
- On the side of the parking lot, smoke escapes through an open window at the end of the second floor of building 4.
- At the back, smoke comes out of the opening in the rear wall of building 7. That opening is closed off with a wooden board, but the top is still one quarter open.
- And finally, on Saint Barbara Street smoke is pouring out through the three windows high in the wall of building 8.

The smoke flows slowly out of these last three places and then the plume drifts toward the residential block (building B) behind the old Delhaize. The current is so calm and the volumes so limited that the smoke dilutes on the spot and vanishes into thin air. There is a light breeze on the front of the building on Coal Mine Lane. Because of this wind direction, it simply was not possible for me to notice any smoke when I coincidentally passed by the building a few minutes earlier. The wind blew the smoke in the opposite direction, away from Coal Mine Lane.

While I am conducting my initial size-up, I find the opening in the rear wall of building 7, through which I will send in the team of Heusden-Zolder later that night. Most likely that opening allowed the owner to get the reusable material of somewhat inconvenient size easily out of the building. At that entrance I can hear the fire crackling. The sound is similar to the crackling you hear during live-fire trainings at the fire academy, when one or more chipboards are fueling the flames. I am not the only person who heard that noise. The owner of the newspaper shop was startled awake by it, but instead of a fire, he thinks there are burglars in his shop. In the past he has received such unwanted nightly visitors. But this time, the burglars must have shown up in large numbers, for they are a bit noisy.

During one of my tours around the building—maybe even the first—a resident of one of the flats of the apartment block behind the supermarket (building B) calls out to me from his balcony: "Is that smoke normal?" I reply: "Yes, there is fire at the front of the building and all the smoke generated by that fire is gently being pushed out through all possible openings." How wrongly can one assess a situation. At that time, I had the impression that the situation was all in all not that bad. The fire we attended to earlier that year in the shop of the Total gas station was a lot more threatening. This time around the color of the smoke is not that bad. It is rather greyish with shades of black. Also, the slow speed of the current is not worrisome. In fact, I am pleased that the building is as leaky as a sieve. All the smoke and heat can escape. A flashover is not possible, because the smoke can get away. Nor do I need to worry about a backdraft. More than enough oxygen can enter the building from various sides. The fire seems to be a textbook example of what is known in the fire service as a ventilated fire development. The amount of fuel determines how long and how intense the fire is going to burn. The strategy is from the onset that we are going to extinguish the fire(s) inside the building and, when our job is done, the building will still be standing.

In the middle of my size-up around the building, I pause for a moment in front of the newspaper shop at the corner of Coal Mine Lane and Saint Barbara Street (building A). The only thing I can see is the upper part of the blank wall of the newspaper shop that towers high above the flat roof of building 1. The newspaper shop appears to be separated from the supermarket by a firewall. That is already one worry less. After the fire, however, it turns out that there is a window in the firewall.

I also do not need to worry about the residential block at the back of the supermarket (building B). On Saint Barbara Street I am looking at the place where the wall of the cold storage of the supermarket (building 8) joins the wall of the adjacent apartment building. A narrow gap of several centimeters separates both buildings. The distance between them gets bigger as you descend further toward the parking lot, as the residential building forms an oblique corner with Saint Barbara Street. I do not have to worry about this building either. The outer wall of the residential block has a fire resistance of at least 1 hour, and the seat of the fire is not at the back of the supermarket but meters away at the front. For sure it is not going to take us an hour to put out the fire. There is no need to evacuate the residential block. By the time everybody has left the building, the fire will already have long been extinguished.

At no time before the dramatic change in the scenario did I think that the fire at the front of building 3 had sufficient strength to affect the roof of the adjacent retail extension (building 6), much less that that roof would turn into a fire load a thousand times bigger than the fire load inside the empty supermarket.

On my arrival at Coal Mine Lane, two police vans have already closed off the road in front of the shop. Two colleagues in the radio room of the fire station of Beringen ask me whether they need to come to the fireground with the signalization vehicle. That vehicle, loaded with traffic cones and various kinds of signal lights, may stay in the fire station. I do not need it at the scene, but I ask my colleagues to remain on standby in the fire station.

When I return to the front after my first size-up, my colleagues have already knocked down the soaring flames. Every firefighter is familiar with this effect that water has on a raging fire. But on Saint Barbara Street, the smoke continues to flow from the windows of the former cold storage (building 8). At around 2:23 a.m., the second fire truck of Beringen arrives on scene, together with our first fire tanker truck. Before I am back at the front, the warrant officer of the second fire truck has already spoken with the owner of the newspaper shop. The owner leads him to the former cold storage of the supermarket (building 8) and points to the smoke flowing from the three windows. He tells him that the white door gives access to the supermarket. The door is broken open.

When I say to the warrant officer that there must be a second seat of fire inside the building because the smoke keeps on coming, he leads me to that white exterior door of building 8. Here we are going to deploy a second line of attack. I ask him if this will work. The look in his eyes says it all: Why should it not work? He is instructed to locate and extinguish any additional fires inside the building with his team. I return to the front side. Immediately after the second team has entered the building via Saint Barbara Street, the warrant officer reports by radio that they have found a kitchen fire. The team is instructed to extinguish that fire. My guess proves to be correct: we are dealing with arson and multiple seats of fire. Two of them have already been found. During the course of the incident, I share that information with a police officer. I confirm to him that he may inform the public prosecutor that the fire is a case of arson.

I myself did not ask the fire truck of Heusden-Zolder for backup. While I am standing in the parking lot, the dispatcher informs me that they have dispatched a fire truck of Heusden-Zolder to Beringen by mistake. It is 2:36 a.m. At the same time, a house has caught fire in the city of Sint-Truiden, and the officer in charge of that incident fears that the adjacent building might catch fire. He requests an additional fire truck. Since the reform of the fire service by the Belgian government in 2015, the fire stations of Beringen, Borgloon, Hasselt, Herk-de-Stad, Heusden-Zolder, Sint-Truiden, Tessenderlo, and Tongeren have been brought together as the Fire and Rescue Service of South-West Limburg. The various fire stations are dispatched centrally from the fire station of Hasselt. "A fire truck for Beringen has been beeped by mistake," says the dispatcher. "Those colleagues are already on their way. Should you not need them, just send them back." At that moment I am standing all alone in the parking lot to the left of the building. I realize that this is an important turning point. I have always been the fire officer who sends back unnecessary material as soon as possible, but now I ask for a time out. I need to think before deciding. The intervention is running like clockwork and it has not occurred to me to request an additional fire truck. The few remaining patches of fire inside the building are almost extinguished.

In my opinion, the fire truck of Heusden-Zolder is not going to make any difference. By the time it has arrived, the kitchen fire will have long been extinguished. But this time I act out of character and decide to let the fire truck drive through. The various incidents of large-scale action training come to mind. However you may twist and turn it, the supermarket is a large building. At present only the front side and the right side are covered. I have got zero extinguishing capability on the side of the parking lot and at the back. Should the event take an unexpected turn, for example, an additional (small) seat of fire at the back, then it is better to have something standing ready there. The fire truck may come through, I decide, but they do not have to rush to the scene. There is no reason to take unnecessary risks. I ask them to approach Coal Mine Lane from the side of the village Beverlo because I want to deploy them at the end of the parking lot near the corner of building 7.

After my radio transmission to the dispatcher, it dawns on me that I still have two men on standby in the fire station of Beringen. They usually sit in the radio room, so they must have heard the message that the colleagues of Heusden-Zolder are on their way to the fireground. To avoid a conflict in my own fire station, I let those men also come to the fireground with our large fire tanker truck. They also do not have to rush to the scene and must approach from the side of Beverlo. At this point I was sure that I was never going to need the 12,000 liters of water in that fire tanker truck, but I wanted to avoid explaining to them after the incident why they were not allowed to come through in support but the colleagues of Heusden-Zolder were. Thanks to the dispatcher, I draw up a plan B for an incident that I was mentally already more or less wrapping up. The irony of fate wants that the victims come from a fire truck that mistakenly was dispatched to Beringen and that I let drive through.

I continue my tour around the building to inform the colleagues at the front and on Saint Barbara Street that a fire truck of Heusden-Zolder is on the way to offer us support. At the front, a firefighter points out to me that the ladder truck has already done its job. For me this is the perfect time to let the ladder inspect the roof of the shop. Earlier that year at the fire in the shop of the Total gas station, I did exactly the same thing in my search for the seat of the fire. The firefighter must ensure that the ladder package is fully extended so that the colleagues in the basket can check for signs of fire on the roof. During one of my later passages, the firefighter reports that there are no visible signs of fire on the roof.

On Saint Barbara Street, the warrant officer appears to be not inside the building but sitting outside on the pavement. That surprises me. On the basis of the radio traffic, I assumed he had entered the building together with his team. Apparently he did not. Two firefighters have traversed building 8 and are now working inside the store extension (building 6). From one of the door openings between these two buildings, a third colleague keeps a close eye on them. In his turn, the warrant officer focuses on the firefighter standing in that doorway and the smoke that keeps on flowing from the windows of the cold storage.

When I am at the corner of Saint Barbara Street and Coal Mine Lane on my way back to the front, I spot a fire truck approaching from the side of Beringen. That must be our colleagues of Heusden-Zolder, although I had asked them to approach the fireground from the opposite direction. That is no problem. I await them in front of the newspaper shop. The fire truck arrives around 2:45 a.m., and I say to them: "Okay guys, you are on the wrong side. I want you on the other side of the building." Coal Mine Lane is packed with fire trucks (ladder truck, fire truck, and fire tanker truck) and there are hoses everywhere. I propose that the colleagues reverse and drive past the casino to Beverlo Road and then take a right turn onto Coal Mine Lane. They reverse, turn, and are off again.

I return to the warrant officer on Saint Barbara Street to inform him that our colleagues from Heusden-Zolder have arrived and that I am going to deploy them at the back of the supermarket. When I am back at the front of the building, I see the same fire truck approaching again, but this time from the other direction. I await them at the corner of Coal Mine Lane and the parking lot. I walk across the parking lot in front of the fire truck to the place where I want them to stop: the corner parking lot and back of the building. The vehicle stops and my colleagues get out in full gear. With the warrant officer and a few of his colleagues in my wake, I walk to the entrance I want them to take: the opening in the rear wall of building 6. I brief them on the situation and tell them they need to search the building for additional fires and extinguish them.

The entrance is a narrow opening. It is boarded up with a wooden board that is more than man-high, but at the top it is still a quarter open and through that opening smoke flows gently out. I tell the team not to go too far inside. Just then, the warrant officer on Saint Barbara Street radios that the kitchen fire has been extinguished. For them the only thing left to do is damp down. I share this information with the colleagues of Heusden-Zolder. At that moment, I think to myself: "Well guys, you have bad luck. All the fun is out of the incident. The teams of Beringen have already extinguished all fires." I end by wishing them good luck and I give one of them a pat on the back.

I leave again and walk to the front of the building and Saint Barbara Street to inform the warrant officers in charge there that the team of Heusden-Zolder has been deployed at the backside. While I cross the parking lot, I notice that still the same amount of the ever-present smoke is flowing slowly out of the last window of the second floor of building 4. In the meantime, two fires have already been extinguished. By now one might expect the color of the smoke to change or, even better, the outflow to come to a stop. None of that, it does not change. Is there yet another seat of the fire somewhere in the building? I walk to the front and ask the pump operator to radio the crews inside that the smoke pattern outside still has not changed. All firefighters inside the building are instructed to be extra careful. I leave for Saint Barbara Street but immediately return to the pump operator of the fire truck at the front. He must radio that message straight away. Again I head for Saint Barbara Street, but this time I hear the message echoing throughout the building.

I do not get far in the direction of Saint Barbara Street. The firefighter who earlier that night pointed out to me that the ladder truck had already stopped extinguishing now informs me that we are running low on cylinders for our breathing apparatus. I suggest that he take the command vehicle and go fetch the remaining cylinders in our fire station. In addition to that, he proposes to also request the breathing apparatus module of Hasselt in support. After the fire, that module has to come to Beringen anyway to replenish our stock, so we might just as well let it come now to the fireground. The dispatcher is instructed to send the breathing apparatus module to Beringen, and the firefighter himself drives to our fire station to gather the remaining cylinders.

A bit later some colleagues are concerned about the fire on the first floor. They fear it might spread to the adjacent newspaper shop (building A). I find that unlikely, as there is not only more than 15 meters (buildings 1 and 2) between building 3 and the newspaper shop but also a firewall. The warrant officer in charge of the operations on the front side knows that there is a staircase behind the inner wall of the family house (building 2). He used it to get to the attic when that caught fire on Christmas night 2018. As the ladder truck cannot reach the fire on the first floor, he is instructed to go upstairs and extinguish it. Together with another firefighter, he is gearing up to head to the first floor.

I continue on my way to Saint Barbara Street and inform the warrant officer of that flank that the team of Heusden-Zolder is on site. On their side everything seems to be under control. His team has extinguished the kitchen fire and now they are damping down any remaining pockets of fire and turning over the debris. When I walk alongside the front of the building, I notice that the fire on the ground floor is picking up again. The last thing I want is for the team on the first floor to get into trouble because that patch of fire on the ground floor has got out of control. I order the pump operator to redraw the team on the first floor and instruct them to put out the fire on the ground floor. The pump operator passes on my order, but then he adds that the backup team can just as well put out that fire. That is fine with me as long as all the fire on the ground floor is out before we send colleagues to the first floor. Talks afterward revealed that the team on the first floor did come down, but they immediately climbed the same staircase again. When they went up for the second time, the conditions in the staircase were completely different. During their first climb there was no heat radiation whatsoever. During the second climb it was sweltering hot. Wisely they decided to retrace their steps.

As I am going to the back of the supermarket, it is approaching 3:00 a.m. A crew member of the fire truck of Heusden-Zolder addresses me in the parking lot. They are still struggling to ensure a constant water supply for their fire truck, and for this reason his crew is not cleared to enter the building. I agree with him: we have to find water. Exactly at that time, the second fire tanker truck of Beringen, carrying no less than 12,000 liters of water, turns into the parking lot. Talk about timing! I say to my colleague: "Your water has arrived." Problem solved. I proceed to the entrance at the backside of building 7. To my surprise, the colleagues of Heusden-Zolder have already entered the building. That is strange; I was just informed that they would not engage until the water supply of their fire truck had been secured. Apparently they thought otherwise. I leave again, but then I retrace my steps and go back to the opening in the rear wall. Only now does it dawn on me that the crackling sound behind the wall has disappeared. Indeed, it is quiet inside the building, completely silent. The fire has been extinguished. We have done well. I inform the dispatcher that the team of Heusden-Zolder has entered the building. Then I walk to the pump operator to ask for a situation report of his team inside. In my opinion, it is time the condition inside the supermarket is clarified once and for all (fig. 2).

6. SUNDAY, AUGUST 11, 2019, 3:08 A.M.

Before I have reached the pump operator of the fire truck of Heusden-Zolder, I hear someone shouting loud and clear over the radio: "Mayday Beringen!" I accelerate and when I have reached the pump operator, I say, "We have a Mayday! We have a Mayday!" after which the evacuation horns of the various fire trucks start to resound around the building. I rush to Saint Barbara Street. As always I do the full tour around the building and run across the parking lot. The left side of the building is still as dark as when the operation started earlier that night. In response to the Mayday, I order the dispatcher to send an additional fire truck for backup.

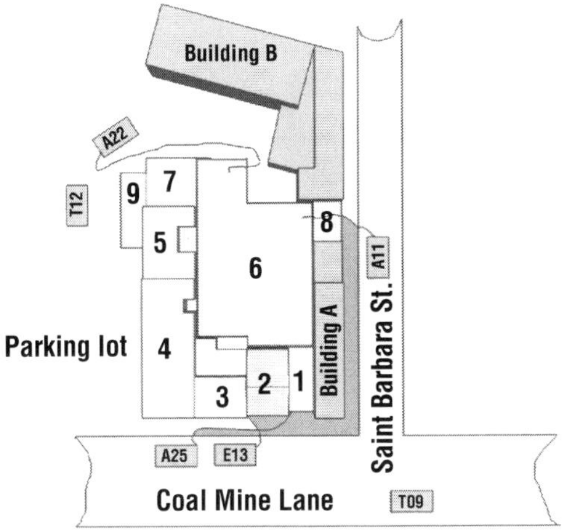

Figure 2. Position of the fire trucks and deployment of the teams
Note. Floor plan by Toon Dumoulin

A 25	Fire truck 1 Beringen	Team 1	Coal Mine Lane
E 13	Ladder truck Beringen	Team 1	Coal Mine Lane
T 09	Fire tanker truck 1 Beringen	Team 1	Coal Mine Lane
A 11	Fire truck 2 Beringen	Team 2	Saint Barbara Street
A 22	Fire truck Heusden-Zolder	Team 3	Back of parking lot
T 12	Fire tanker truck 2 Beringen	Team 3	Back of parking lot

I cannot imagine a Mayday from the team working on the front side. That part is, of all the facades, the most open one. The Mayday is probably called by the team on Saint Barbara Street. During my passage at the front, I rapidly check my assumption with the pump operator of the fire truck standing there. When I arrive on Saint Barbara Street, the missing colleague of Beringen has already been pulled out of the building by his colleagues. He is lying on the ground and I ask if he needs an ambulance. His answer is short but clear: "No."

Suddenly I see yet another firefighter lying on the ground a little further down the road. His protective clothing is evaporating heavily. I walk over to him. He is still wearing his breathing apparatus and the full-face mask covers up most of his face. I do not recognize the face I am seeing through the visor. I ask him if he is from Beringen or Heusden. He replies: "Heusden."

That is strange, I think to myself. I am now on the side of the team of Beringen. How can someone of the team of Heusden end up on this side? At that moment I am still not aware of the narrow passage at the back of the supermarket. I suppose the colleague must be confused by what he has just experienced.

I go to the warrant officer on Saint Barbara Street to check whether everyone on his team is out of the building. He answers in the affirmative. And I have got two firefighters lying on the street. That must be correct. His team is outside. The response "Heusden" must be a mistake, and with great relief I radio to the dispatcher that the Mayday is solved.

My feeling is one of immense relief: "Phew! That was a close call. Fortune was on your side tonight. You almost lost one of your team members. Luckily everyone is now outside." A quick glance in building 8 suffices to conclude that the conditions inside have changed for the worse. Pitch black smoke banks down to floor level and the visibility inside is reduced to zero. Turbulent smoke is billowing out of each opening.

Barely recovered from the first Mayday, I now hear a soft, calm voice saying over the radio: "Mayday Heusden, Mayday Heusden." I think I heard it on my radio and for that reason I will later try to communicate directly with my colleagues of Heusden-Zolder. Only days later did I realize that I could not have heard the Mayday of Heusden on my radio. The teams inside use a different radio channel (DMO or Direct Mode Operation) from the incident commander (TMO or Trunked Mode Operation). That is why all my communication with the teams inside was done via the pump operator of the fire truck. He wears the radio jacket with one radio tuned to DMO and another to TMO and holds the function of radio operator. I must have heard the Mayday on the radio of one of the colleagues of Beringen standing next to me.

After the Mayday of Heusden, a feeling of incomprehension prevails. What are those colleagues still doing inside the building? A Mayday means everyone has to leave the building, or does it not? And do they not have to evacuate right away? But why did they not do that? Could they have missed the Mayday of the team of Beringen? Once more I return to the back of the building. On my way, the answer of the second victim on

Saint Barbara Street sinks in. I have a firefighter of Heusden on the side of Beringen. Maybe the second Mayday is a mistake, and the firefighter of Heusden on Saint Barbara Street might well be the reason. They are missing one of their colleagues. Or is this all just wishful thinking on my part? After all, I do not want another Mayday with colleagues from Heusden or anyone else in that building packed with pitch black smoke. The rushed current of the dark rolling smoke plumes does not bode well. A flashover is imminent.

I need to find the crew of the fire truck of Heusden-Zolder. Only they can provide answers. I have only taken a few steps down Saint Barbara Street when, via the radio, the dispatcher informs me that the automatic smoke alarm has been triggered in the newspaper shop (building A). At that moment I still had to pass the corner of Saint Barbara Street and Coal Mine Lane. Until then, everything had been fine in that shop. At regular intervals I checked the conditions inside with the shopkeeper, who was standing on the pavement. Apparently some smoke has now managed to enter the store through an opening. I look for the shopkeeper but I do not find him right away. With the second Mayday claiming all my attention, I have no time to waste. I order the colleagues at the front to go find the owner of the shop. They have to ask him to open the shop door so that they can pressurize the newspaper shop with a positive pressure fan, preventing further intrusion of smoke. Later that night, during my endless wandering around the burning building, I spot a switched-off positive pressure fan standing in front of the closed door to the newspaper shop.

During my umpteenth passage along the front side of the building and across the parking lot, it is quite obvious that the conditions inside the supermarket have changed dramatically. The building is no longer as dark as the night; inside everything is brightly lit. Most of the shop windows on the side of the parking lot were covered up with metal or wooden boards. A black plastic foil was stuck onto the inside of the shop windows that were not covered up. Now an immense wall of flames lights up through the countless bubbles and holes in that foil. I try to establish radio contact with the colleagues of Heusden-Zolder inside the building. All my attempts are in vain. They do not respond. They cannot hear me on the TMO channel of my radio, but the reinforcements already on the way from the fire stations of Heusden-Zolder and Tessenderlo hear me: "Team Heusden, are you okay? You have had a flashover." According to one of the colleagues on board one of the responding fire trucks, I have repeated my question four times, and panic can be heard in my voice.

I go to the corner of the parking lot and back of the building, where the fire truck of Heusden-Zolder stands. There I find the colleague with whom I discussed the water supply for their fire truck just a moment ago. He immediately crushes my last sliver of hope: two of our colleagues are trapped inside the building! I speed to the place where I have sent them in. Entering through that opening is no longer possible. At the entrance I am facing flames, flames, and even more flames. They are all more than knee high. Yellow flames, thick and dense. Individual flames can barely be distinguished. They form one monolithic block, as if a forklift has left a pallet of fire at the entrance, but in this case the whole building is stacked with burning pallets. My guess is that we have lost those colleagues. No one comes out alive from such a sea of fire, and I cannot send anyone in to offer help. Furthermore, I do not have much time to resolve the predicament we are in.

After this unexpected turn of events, it is literally hell on earth at the back of the supermarket. Fire, fire, fire. Fire everywhere. Despite the hopeless situation and against my better judgment, I must try to get my guys out of the building. Or at the very least I must try to get the exit clear of fire so they have a chance to get out of there. I need to get extra extinguishing power at the back of the building, and I need it fast.

I run across the parking lot to the front to find firefighters that I can deploy at the rear. Finding somebody is not easy. The first person I address replies that he is out of air. He has to replace the cylinder of his breathing apparatus. This is going to take too long. By then all help will be too late. Instead of asking, I now simply point out the colleagues that have to follow me. With that team I will deploy an additional handline through the narrow passage on Saint Barbara Street and attempt to reach the entrance at the back of the building.

In the meantime, a member of the fire truck of Heusden-Zolder has pointed out to me that there is a narrow passage at the back of the building that leads to the adjacent street. He used it earlier to drop his colleague on Saint Barbara Street and then he immediately rushed back to the entrance in the hope he could evacuate his two other colleagues in the same way. No one from the team of Beringen had seen him do this. That is why I could not figure out the presence of a colleague of Heusden on the side of Beringen. On Saint Barbara Street we extend the extinguishing line with an additional hose, and with that charged handline we traverse through the narrow passage and hurry to the entrance at the back. Meanwhile the whole corner of Saint Barbara Street and the back of the shop has turned into a gigantic wall of flames. A part of the roof has already collapsed. With the additional team I cannot reach the entrance at the back. Our efforts prove unsuccessful. The huge, bright wall of fire we encounter on our path is simply too overwhelming.

I return through the narrow passage to Saint Barbara Street, where a police officer is standing on the pavement near the second fire truck of Beringen. It is probably the same police officer I told earlier that the fire is a case of arson. I inform him that I do not think we are going to get our two missing colleagues out of the building alive. He may already notify the legal authorities.

I go back to the front of the building, where I meet a fellow fire officer of Beringen. He is responding to the automatic fire alarm in the newspaper shop. To the best of my ability, I brief him on the situation while we are walking to the back of the building. There I hand over command and radio the transfer of command to the dispatcher. I just can't anymore. In the meantime, the pattern of the flames has changed at the back of the building. Now the all-consuming flames come out of the building in waves. One wave after the other rolls toward me, and they just keep on coming.

My brain shuts down. Enough is enough. The stress level has risen too high. I am overwhelmed by the horror of the situation. This has consequences for this account, as there are more and more gaps in my memory. I have got fragments of scenes and images in abundance, but their coherence and location in time are fuzzy.

After handing over command, I start to walk around the burning building like a zombie. From that moment on my search for explanations starts.

What went wrong? But most importantly: *What have I done wrong?* Which signal or signals did I miss? I wander and wander and wander only to get hopelessly lost in my own head. It is going to take me days, weeks, even months to get things straightened out.

During this endless wandering around the building, my two apparitions are the most remarkable experience. I do not have a better word for those experiences than this highly charged religious term. At a certain moment, I recognize one of my colleagues of Heusden-Zolder at the backside of the building because he is not wearing his fire helmet. When our eyes meet, an indescribable ease and calm descend over me. Afterward I found out that he was assembling the remaining crew of the fire truck of Heusden-Zolder to get them transferred to their fire station. A few minutes later I recognize a warrant officer from Tessenderlo in the parking lot. He is also not wearing his fire helmet. Again our eyes meet and again the same indescribable ease and calm descend over me. Only then do I fully realize that the requested reinforcements have arrived. With the help of the colleagues of Heusden-Zolder and Tessenderlo, everything will be sorted out soon.

This indescribable sense of relief stands in stark contrast to my reaction when I see fellow fire officers for the first time. Then fear and even more shame prevail. I address the fire chief of our Fire and Rescue Service. I tell him I am responsible for the whole mess. When he points out that he as fire chief is ultimately responsible for everything, he somewhat relieves the heavy weight that burdens my shoulders but even more my mind. For a short period of time I get a bit more breathing space.

Then I see a fellow officer of the Fire and Rescue Service of North-Limburg standing in the parking lot. "Fellow officer" is a wrong description, because over time he has become a friend of mine. Even in a firefighting suit and with his fire helmet on, I recognize him on the spot, but . . . I cannot face him. I want the earth to swallow me. I cannot think of anything better than to bury my head even deeper in my fire coat and ignore him completely. I quickly walk away, secretly hoping that he has not recognized me.

On one of my later tours around the building, he is standing there again. I realize I cannot ignore him forever. Sooner or later I will have to face him, and for that, any time is equally good or bad. I take a deep breath, gather all my courage, and address him. For the first time I tell my story to a fellow fire officer. When I am finished, he replies: "The way you tell it, I would have done exactly the same thing." The colleague that I at first did not dare to face now throws me a lifeline that I could never have dreamed of. Over the following days, several other people will tell me exactly the same thing.

When, hours later, the missing colleagues are finally located inside the building, I am at the front. I hear the message over the radio. I cannot and do not want to go to the side of the parking lot. Seeing the colleagues that I have sent in earlier that night being carried out of the building into the parking lot is something I cannot handle. That image would have led to a complete breakdown and remained upon my retina even longer than the last image I have when I leave the fireground later that morning: two tents of the fire service in front of the left outer wall of the burned-out supermarket. I know all too well what is inside those tents.

In the early hours of the morning, a sector commander sees me while I am still wandering around the building. He is of the opinion that I should also go to the fire station of Heusden-Zolder. About what I am to do or expected to do there, I am completely left in the dark. The entire crew of the fire truck of Heusden-Zolder has already been taken there. The reinforcements from Heusden-Zolder and Tessenderlo that arrived later at the fireground are also or almost there to attend a debriefing. A colleague of Beringen drives me to Heusden-Zolder. Upon arrival in the fire station a woman addresses me. Last night she was on duty in the control room of the emergency number 1-1-2. To judge by the radio communication she was able to follow, I have acted correctly. That is a relief. The cafeteria is crowded. I cannot find any peace at all. On the contrary, I start to pace up and down the cafeteria. For some reason, moving around helps me clear my head. And I have more than enough to break my head over: *What have I done wrong?* To avoid making everyone nervous, I leave the cafeteria and start to pace up and down the long corridor next to the cafeteria.

Suddenly it dawns on me that I still have to tell my wife what has happened. I ask if I can make a phone call. I do not have my smartphone on me; I had given it to Grace Saturday evening when she left the sports hall. But I cannot call her because I do not know her number. The only phone number I know by heart is that of my parents' landline. My mother answers the phone. I tell her what has happened and ask her to call Grace. Then I come across a fellow fire officer in the corridor who says I have to call my wife. I answer that I cannot do that because I do not know her number. Her number is in my smartphone, and I do not have my phone. It is with her. "But then you can call her on your own phone," he quite rightly points out. I call my number on his smartphone. My wife answers. I go outside and, in a haze of anxiety, wander around the neighborhood of the fire station. With tears in my eyes I tell her what has happened and what I have done: I have sent colleagues into a building on fire and two firefighters have lost their lives. A third one had to be rushed to the hospital with severe burns. How hard all of this must be for their wives and children!

In the fire station of Heusden-Zolder I am asked to give a debriefing. I just have to tell what happened last night. Why not? I vividly remember the trauma psychologist and the classroom in which the debriefing took place. I can also see myself get up from my chair to take the floor, but then it stops. I have no recollection whatsoever of the things I said or saw at the debriefing. My mind is a complete blank.

The debriefing was for all operational personnel—firefighters and paramedics—of the fire stations of Heusden-Zolder and Tessenderlo deployed at the fireground. As a matter of fact, the warrant officer of Tessenderlo who at the incident ground made me experience something similar to an apparition stood all the time next to me and even held me. But I have no memory of his presence. To this day I cannot remember any of the faces I saw at the debriefing. I was completely dazed. Some people say I just stood there jabbering away to the white board with my back turned to the audience. According to another colleague I clearly was drawn deeper and deeper back into the horrific incident as the story unfolded.

After the debriefing, my colleague takes me to the fire station of Beringen for yet another debriefing. There are not many fellow firefighters there. Most of them have returned home, having already been debriefed on site. Again I give my account of the events, then I go home. Before leaving the fire station, I call the dispatcher in Hasselt. I recognize the voice of a warrant officer of Beringen on the other end of the line, but I cannot come up with his name that is on the tip of my tongue. I ask him to change my status in the paging system to "not available" for the rest of the year. I no longer see myself functioning as a fireground commander for the remainder of 2019. In fact, I never see myself taking up that job again. Alone and empty, I drive home.

7. SUNDAY, AUGUST 11, 2019, 11:00 A.M.

Nothing any longer feels real. Everything is different and feels strange. I am living in a daze.

2

LESSONS FROM THE FIRE

In February 2020, the Federal Public Service Home Affairs released the incident report on the fire of Beringen. One of the conclusions reads as follows: "The different teams were deployed in a logical manner by the officer." However, the outcome was dramatic. Two firefighters lost their lives and a third colleague suffered severe burn injuries. How is it possible that despite my logical and rational approach, the operation went so horribly wrong? In this chapter I try to clarify that paradox.

1. AFTERWARD YOU KNOW EVERYTHING BEFOREHAND

Now that you are reading these lines in the comfort of your office or at home on the couch, the fateful outcome of the incident is in the back of your mind. Already from the start, you know that things will end dramatically and firefighters are going to lose their lives. For every reader, hindsight bias and retrospective distortion are the most difficult hurdles to overcome while reading this chapter, as you are familiar with the outcome in advance.

Hindsight bias is our innate tendency to vastly overestimate our ability to predict an event after the facts: "I told you so that sooner or later it would go all wrong with interior firefighting. I knew it all along: far too dangerous." The delusion of retrospect.

As with all incidents, the fire service can only assess this incident after the event. But when we are looking in the rearview mirror, we are hardly aware that with each glance back, we change the sequence of facts of that incident—or any major event of our life, for that matter (e.g., your wedding day or the day of birth of your children). In retrospect, the sequence of events is always more logical and less random than actually was the case. It creates the illusion of posterior predictability as we are incapable of ignoring the outcome. This process is known as *retrospective distortion*. All these adjustments are the result of the way memories are formed and stored in our brain.

The process of how our brain creates a memory is similar to the working of the word processing program on your computer. The *hippocampus* plays an important role in the creation of a memory. This brain region of just a few cubic centimeters owes its Latin name to its peculiar shape: it (vaguely) resembles a seahorse. The devastating effects of Alzheimer's disease begin with neurodegeneration in this brain area. When the brain cells of the hippocampus die, the result is loss of memory. The chemical storage of a memory, on the other hand, takes place in the *neocortex* or the pleated outer layer that surrounds the cerebral hemispheres, where the various parts of a scene are recorded in proteins in different brain regions. That is why in the early stage of Alzheimer's a person can still talk vividly about things that happened a long time ago but will simply forget the just finished conversation and end up repeating the same story or asking the same question over and over again. Each time you retrieve a memory, all those components stored in various brain areas travel back to the working memory at lightning speed. There they are highly vulnerable to reshaping by an adjusted version of the events as, through *egocentric bias*, your role in the events becomes more prominent than actually was the case. For example, the catfish that a fisherman managed to catch a few weeks ago is always getting bigger and bigger and the fight to get it onto land becomes more and more heroic each time he tells the story.

Later in this work you will come across a diversity of brain regions (e.g., amygdala and prefrontal cortex). A brain area never functions alone. They are all connected with each other via several routes and partially they perform their functions together. Sometimes there is overlap in function, but then each brain region fulfills a highly specific task within that overlapping function. You can compare the different brain areas to members of a family. A family functions as a single cooperating unit, but sometimes the various members can sharply disagree about what to do and struggle to make a decision.

The Lebanese American essayist N. N. Taleb summarizes the impact that hindsight bias and retrospective distortion have on our illusion of memory as follows:

> We will tend to more easily remember those facts from our past that fit a narrative, while we tend to neglect others that do not appear to play a causal role in that narrative. Consider that we recall events in our memory all the while knowing the answer of what happened subsequently. It is literally impossible to ignore posterior information when solving a problem. This simple inability to remember not the true sequence of events but a reconstructed one will make history appear in hindsight to be far more explainable than it actually was—or is.[1]

Western historiography abounds with examples. In every book on the world wars, you will read unequivocally that the Treaty of Versailles, which in 1919 put an end to World War I, was the trigger of World War II. As unavoidable as this is now claimed to be in our history books, so entirely unthinkable was the whole idea for the negotiators of the Allied powers at that time. The content of the treaty would indisputably have been different had they been aware that in less than 35 years it would inevitably lead to a new world war, one that would turn out to be even more bloody. But would there ever have been a World War II if Hitler had never been born or had been killed during World War I?

According to the Israeli historian Y. N. Harari,

> History cannot be explained deterministically and it cannot be predicted because it is chaotic. So many forces are at work and their interactions are so complex that extremely small variations in the strength of the forces and the way they interact produce huge differences in outcomes.[2]

Just like in the career of a successful businessperson, scientist, or top athlete, in the explanations that we come up with in retrospect, we overtly underestimate the role of luck. We seem to be guilty of *hineininterpretierung* (overinterpretion) of the past based on the eventual outcome because we find it hard to accept that some things happen by accident while others do not.

Now back to Beringen and the course of events during the night of the fire. As you are working your way through this book, your focus will unconsciously be on those moments where a slightly different decision would have produced another—even a better—outcome. For example, why send firefighters into a large, empty building that should have been demolished years ago? There was nothing to gain, only to lose. I fully understand that focus. But you have that focus because you know that there are going to be fatalities. That focus was completely absent the night of the incident. At the start there was not a single element that hinted at the slightest possibility that we might incur casualties. And in Belgium, it is also not a rule that the fire service may not mount an interior attack when attending to a fire in a large, empty building.

To temper that innate bias somewhat, read this work as if it were Saturday evening, August 10, 2019. Imagine you are sitting at home enjoying a lovely summer evening. You have no idea of the drama that is about to unfold over the next few hours. The factual account of the course of events can be found in the incident report released by the Federal Public Service Home Affairs in February 2020. I summarize it briefly:

> In one of the retail spaces of a complex building of roughly 2,500 m^2, a fire started in the worst possible place. The initial seat of the fire is located near where the concrete vaults of the ceiling of the retail space at the front of the building (building 3) connect to the wooden roof structure of a 830 m^2 extension (building 6) added later. The fire manages to get into the roof structure. There, fuel is for taking (including, among others, wood, plenty of insulation, and two layers of tar). The castellated beams—that is, steel beams with a repeating pattern of hexagonal holes—in the roof make it easy for the heat and the fire to advance through the ceiling toward the rear of the building. The back wall of the

> store is clad with chipboards. The high temperature inside the ceiling initiates the thermal decomposition of those chipboards. The decomposition gases released through this process of thermal cracking or pyrolysis literally pull the fire out of the ceiling at the back of the building.

The strategy that I have put together that night runs counter to the factual course. When I arrive, I find a blazing fire at the front of a delipidated supermarket. Luckily the fire is not located deep inside the building. The greatest danger can be dealt with by an exterior attack. The supermarket is not unknown to the fire service of Beringen: In 2018 alone, there were three fires in that building. I assume that we are dealing once more with arson. After all, the building was vacated years ago and all utilities (electricity and gas) had been cut off a long time ago. I keep in mind that there might be multiple seats of fire in the building. That is not uncommon when you are dealing with arson.

The strategy can be summarized as follows: We locate and extinguish the (seats of) fire, and when our job is done, the supermarket will still be standing. For this reason, I do not evacuate the residential block (building B) situated at the back of the shop. Nor do I expect problems in the newspaper shop (building A), which is built next to the supermarket on the corner of Coal Mine Lane and Saint Barbara Street.

While I conduct my initial size-up around the supermarket, I notice that at several places light black-grey fumes slowly flow from the building. I can explain perfectly what I am seeing. The raging fire at the front of the shop places the whole structure in overpressure. The building is leaking like a sieve, and through all possible openings the smoke produced by the blazing fire is gently pushed out the building. All information points toward a ventilated fire development.

After the fire at the front of the supermarket has been extinguished, the pattern of the smoke does not change. The fumes just keep on flowing slowly from the building. In all likelihood there is another seat of fire inside the building. Along the side of Saint Barbara Street I send in another team of firefighters. Shortly after entering the building, they report to have found a kitchen fire. My assumption of a second fire is confirmed. This second fire is the reason why smoke keeps on coming out of the building. Around 2:50 a.m. I send in a third team via the back of the building. The colleagues of Heusden-Zolder are instructed to check whether there are any remaining seats of fire and, if so, to extinguish them. When I send them in, I receive over the radio the message from the Saint Barbara Street team that they have extinguished the kitchen fire. They start to damp down.

2. IS SEEING BELIEVING, OR BELIEVING SEEING?

Incident command at a fireground is in essence nothing more than information processing and decision-making by the incident commander. Knowing how the brain processes information that enters via the sense organs is not an unnecessary luxury. For an answer to that question, I consult with the Dutch neuropsychiatrist A. Aleman. We all have the impression that our eyes function like a video camera that minutely records everything we see and provides us an accurate and objective representation of the world we live in. All too often we overlook that the real seeing is done inside and through our brain.

Driven by stimuli from the outside world, information enters the brain through the eyes. Upon entry that input is immediately unraveled into small bits and pieces of information (lines, angles, contrast, shape, color, and movement) and transmitted to the higher brain structures in the cerebral cortex for further analysis and interpretation. This part of the trajectory is called *bottom-up processing*. After being processed and enriched in different areas of the cerebral cortex, all those scraps of information have to come together again in order to be integrated into a meaningful whole. Our brain does that reconstruction under the influence of prior experiences, knowledge, expectations, and wishes. This part of the trajectory is the *top-down processing*.

How much of the information our brains process is bottom-up or top-down? According to the Dutch molecular neurobiologist E. van der Zee, our brains add more top-down information than they process bottom-up impulses. This also applies to the brain areas that are directly involved in processing information from the outside world, such as the visual cortex at the back of the brain. Barely 10% of the processed data is sensory input. The other 90% originates from within the brain itself. This concerns mainly internal communication between nerve cells in the visual cortex and information provided by other brain regions.

An online video showing little more than a mask turning around aptly illustrates the impact that top-down information has on our perception.[3] As soon as the concave side of the mask turns forward in the video, you will perceive it as convex. Formed by a lifetime of experience and practice, your brain knows that faces are not hollow and that noses do not stick in. They stick out. Your brain radically adjusts the image and changes the concave inside of the mask into a normal spherical face. In the case of noise or unclear information, such top-down information is most influential. Your brain then complements and adjusts the incoming image or sound as it sees fit with whatever makes the most sense. In our brains we carry a model of the world that we constantly check with our current observations (bottom-up) and past experiences (top-down). The brain as a predictive tool or an anticipation machine is not a reactive but a proactive system that constantly uses the prior knowledge we have acquired. And with everything we do, we are always updating and expanding our inner mental model of the reality we are experiencing.

Perhaps no one is more aware of the importance of experience in processing visual input—or rather, the many problems that arise when lacking that expertise—than Mike May, a Paralympic athlete who excelled in numerous disciplines of blind skiing in the 1984 Winter Games. When he was three, Mike went blind after he got a corrosive substance in his eyes. However, at the age of 46, he was able to partially see again after a cornea transplant. Because he had had no vision for 43 years, his brain was insufficiently trained in quickly and correctly analyzing and interpreting visual input. Therefore, he could not recognize faces, not even that of his own son. Once, after a football game, he offered the ice cream he had bought for his son to another player of the team. That is what you get when the whole team plays in the same shirt! And he stared into the eyes of his wife's friends for such a long period of time, and so intensely, that not only those friends but also his wife felt uncomfortable. Depth perception was also problematic because his brain was insufficiently familiar with the subtle interplay of shadow and movement. Blind, Mike was able to move around smoothly without any assistance. When his vision was restored, he needed the help of a guide dog.

Perception is controlled far more from within the brain than by stimuli from the outside world. In addition, much of this information processing happens outside our conscious awareness. The aforementioned 90% is a conservative estimate. According to some neurologists, no less than 95% or even up to 99% of our brain processes are unconscious. Advertisers make good use of this. With their advertisements they stimulate this unconscious information processing, knowing that in doing so they might succeed in orienting our purchasing behavior toward their products. Advertisements have far more power over you than you are aware of because what you see is what you do, especially when you are not paying close attention to the messages bombarding you. However, conclusive scientific evidence that advertising can subliminally influence consumer behavior is still lacking. When we compare the total amount of brain activity to an iceberg, our stream of conscious thoughts is just the tip. We must not make ourselves any illusions: in our brain, unconscious processing of information and unconscious thinking takes place much more than conscious or rational thinking.

All in all, large amounts of stimuli enter the sensory memory: through the eyes alone, 10 million bits per second. It is simply not possible (and also inadvisable) to process all that sensory input. To save energy, your brain limits itself to processing only what it deems necessary. That selection is made on the basis of attention. Only what you pay attention to or what you are looking for will be processed.

An online video with some basketball players, part of a British road safety advertising campaign, illustrates this aptly.[4] Before the start of the video, you are asked to count the number of passes made by the players wearing white jerseys. Everyone has counted 13 passes at the end of the video. But did you notice anything other than the players?

About half the people fail to see the man dressed up in a bear suit that prominently walks through the shot. Have you seen him? Watch the video again, but this time do not focus on the ball. The bear features so prominently in the center of the screen that now you are wondering how you have missed something so obvious right in front of you. The answer is easy. You were focusing intently on the number of passes. Furthermore, you did not expect to see a moonwalking bear (or a chest-thumping gorilla, should you have found the video of the original experiment). This error of perception is called inattentional blindness or the illusion of attention. You are effectively blind to objects you do not expect to see and you do not pay attention to. This is precisely what illusionists and magicians make good use of. By subtly manipulating and distracting your attention, they surprise you with their most amazing tricks. They know all too well that you can only fully focus your conscious attention on one thing at a time.

At home, parents experience the same with their growing teenagers. Ask your child to do something when they are in the middle of a fight to the death in front of the PlayStation or Xbox. You will get no reaction. If you ask them whether they have heard you, the answer is yes. If you ask them what you have said, they remain silent. Apparently they have heard you, but they did not listen to what you were saying. From personal experience, I know that you are more likely to get a response when you start your question with their name. That is how you draw their attention. Saying someone's name activates a bottom-up attentional process that always works; even at a crowded party your attention will be caught when you suddenly hear someone saying your name over on the other side of the room. This social psychological phenomenon owes its name to this: the cocktail party effect.

In daily life we can only focus our attention on one thing at a time. We solve this by constantly switching our attention between different tasks. This does not mean that you cannot do two things at the same time. Everyone is capable of doing that, although in my experience, women are better at it than men are! You can do several things at a time, for example, driving a car or riding a bike while talking to the person next to you, provided you have internalized those actions and they have become an automatic thinking pattern, such as driving or cycling in our example.

Without noticing, we miss information that is obvious, on the one hand, and on the other, we easily fill in missing bits of information. There are many other factors that contribute to processing perceptions and determine how you interpret information: for example, context and expectations.

In the Ebbinghaus illusion, the circles in the middle of both figures are of identical size, but the size of the surrounding circles gives the impression that the circle in the middle of the right figure is much bigger (fig. 3). And in Müller-Lyer's famous optical illusion, both lines are the same length. Just measure them. The outward arrowheads make the bottom line just look longer (fig. 4).

I described earlier the color of the smoke that slowly escaped from the building in Beringen as light black-grey. That was my perception around 2:00 a.m. against a dark background as I was standing on the pavement in a poorly lit street. In broad daylight my perception of the same smoke would for sure have been a lot darker. Psychologist D. Kahneman already noted that "reference dependence is ubiquitous in sensation and perception."[5]

Expectations can improve your visual acuity. In our minds, the bottom rows of letters on the eye chart at the doctor's office are always illegible because they are printed way too small. Many past experiences have taught

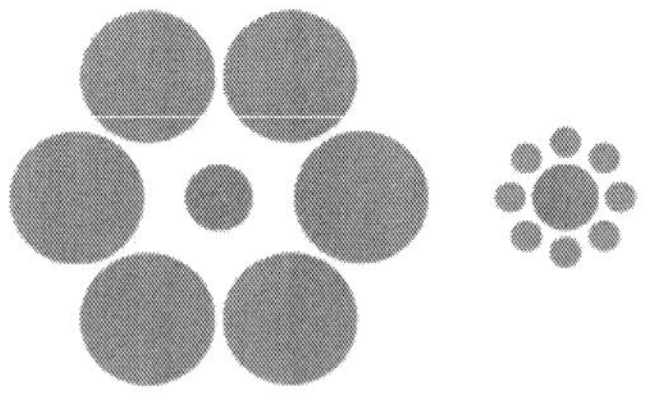

Figure 3. The Ebbinghaus illusion

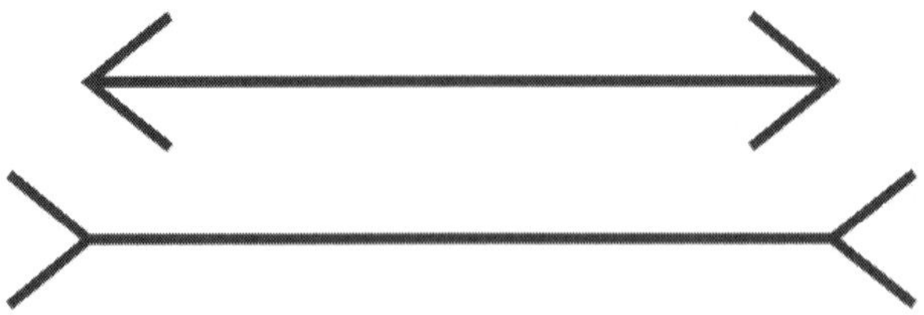

Figure 4. The Müller-Lyer illusion

us that. In a psychological test, the order of the rows on the chart was reversed. The small letters were printed at the top and the large letters at the bottom. As with each eye test, the physician started with the top row. And guess what? The test subjects were able to read much more fine print than in the traditional setup.

All in all, our perception turns out to be all but an exact replica of the reality around us. Or in the words of the Dutch neuropsychiatrist A. Aleman:

> We think we see reality as it is, but in practice we do not see much. And what we see, we perceive through the lens of our experiences and expectations. Our perception is much more an own construction than a faithful reproduction of reality.[6]

3. THINKING ABOUT THINKING

Everyday life is nothing but a succession of decisions, a few big and many small. Most of the decisions we make are good; some are poor and others all wrong. The way we process information and subsequently make judgments has been extensively studied by D. Kahneman. This Israeli American psychologist was awarded the Nobel Prize in Economics in 2002 for his research. His best-selling book *Thinking, Fast and Slow* is one of the most significant books on psychology in recent years. The idea that most of us are overconfident and prone to put far too much faith in our intuitive judgments runs like a common thread through the various articles and books of this distinguished scientist.

Let us turn to him to gain some insight into the mental mechanisms inside our head. A dual mode of thinking, which he calls the automatic System 1 and the effortful System 2, occupies center stage.

System 1 contains all your automatic mental processes. It operates automatically and quickly, with little or no effort at all and no sense of voluntary control. It is your reflexes, intuitions, and automatic thinking patterns.

For me, reflexes and intuitions are different things. You have your intuitive beliefs and I have got mine, but we both share the same innate reflexes.

An intuition is a learned thinking pattern with which you quickly interpret everything that happens around you without thinking too hard. It is in fact the knowledge or the experience that you have acquired in the past and that now directs your behavior automatically in new and unfamiliar circumstances. Broadly speaking, at a fireground the intuitive beliefs of each fire officer will correspond more or less with mine. There is bound to be a lot of overlap, but there will also be some minor differences as our skilled intuitions have been formed by different fires that all have run their own course. And then I do not even mention differences in training, education, and upbringing.

A reflex, on the other hand, is an innate behavioral pattern. For example, we all share the same fight-or-flight response that was first described in 1915 by the American physiologist W. Cannon and to which later "freeze" was added. When we are in danger, we all flee or fight back. It is our natural response to events that are perceived as stressful or frightening. Because in the past this evolutionary adaptation has significantly improved our chances of survival in life-threatening situations, it was inscribed into our DNA over the course of the centuries-long process of natural selection. Thus, the behavioral pattern of the three Fs is now passed on from generation to generation.

An automatic thinking pattern, finally, is a learned ability or a skill that you have acquired at one time in your life through training or schooling that, thanks to prolonged practice or usage, you can now perform without deliberate thought, like reading or cycling.

System 2 of Kahneman contains all the effortful mental activities that require attention and are disrupted when that attention is drawn away. They expend mental energy and are slow. These include, among others, complex calculations (*357 × 23 = ?*), planning (*What are your plans for tomorrow?*), self-control (*I have to stay calm!*), and decision-making (*Are we going to buy this house?*). The operations of System 2 are often associated with the subjective experience of agency, choice, and concentration.

For sake of convenience, I use the terms *automatic* (1) and *conscious mental processes* (2) instead of System 1 and System 2. The numbers in parentheses refer to Kahneman's numbering.

In daily life, both automatic (1) and conscious (2) modes of thinking are active. Depending on the situation, they interact with each other to a greater or lesser extent. In some circumstances there is no interaction at all. To solve a complex computation in your mind is a purely conscious mental process (2). It is unlikely that you intuitively know that 357 × 23 is equal to 8,211. Our fear reflex, on the other hand, is an innate automatic behavioral pattern (1).

Every second of the day automatic mental mechanisms (1) are generating suggestions in the form of an endless stream of impressions, intentions, and feelings. Even if you wanted to, you cannot turn off that stream. The speed of the stream may be superfast, but the quality of the suggested ideas is far from splendid. More than once the suggestions are imperfect as the whole system is riddled with intuitive biases or systematic prejudices.

Your conscious mode of thinking (2) is quite a different story. It is born tired and its basic state is the *standby mode*. When endorsed or rationalized by conscious mechanisms (2) your impressions, feelings, and inclinations are turned into beliefs, attitudes, and intentions.

One of the most important tasks of conscious mental processes (2) is to monitor and maintain control of that endless stream of freewheeling impulses created by the automatic mental mechanisms of System 1. Contrary to what you might expect, conscious mental processes (2) let countless fallacies slip through the cracks. Most of the time, the suggestions of System 1 are adopted with little or no modification. Only a few intentions and intuitions are suppressed. When unconscious mental mechanisms (1) immediately provide a very logical-sounding answer, there is barely any conscious control (2) exercised over it. Together with the proposed conclusion, all arguments that appear to support it are accepted as valid, even when they are unsound. When automatic modes of thinking (1) predominate in decision-making, the conclusion comes first and the arguments follow suit.

Here are two examples. Give the first answer that comes to mind:

> A bat and ball cost $1.10.
> The bat costs $1.00 more than the ball.
> How much does the ball cost?

Like most people, you probably intuitively responded that the ball must cost 10¢. You better check that answer. If the ball costs 10¢ and the bat costs $1.00

more, then the bat costs $1.10, for a grand total of $1.20. The correct price of the ball is 5¢ and the bat costs—at $1.00 more—$1.05, for a grand total of $1.10.[7]

Another example. Fill in the blank:

John's mother has three children.
Their names are May, June, and . . .

Intuitively you fill in *July*. But her third child is of course John.

As stated above, conscious mental processes (2) barely exercise control over very logical-sounding answers that are immediately provided by intuitive thinking (1). It is typically human to quickly and intuitively draw conclusions based on limited information or flimsy evidence. For this behavior Kahneman coined the acronym WYSIATI, which stands for "what you see is all there is."

A good grasp of this mode of thinking is essential. When making an intuitive judgment (1), neither the amount nor the quality of the information counts for much in our subjective confidence. We are overly optimistic about what we think we know and grossly underestimate the vast amount of information that is missing. The scarcer the information, the faster we jump to conclusions. All uncertainties and doubts are suppressed.

What is the subjective confidence in our first impressions or intuitive judgments based on, then, if the amount of information is slight and the quality poor? All that matters is the coherence of the story you have managed to create. If your associative brain succeeds in immediately constructing a consistent and plausible story from the scraps of information available at that moment (*May, June, and . . .*), then the ball is rolling. And that is something automatic mental mechanisms (1) excel in!

Unless the story is soon after negated and rejected as false, the (unconscious) associations it keeps on invoking will only make it appear more consistent. If new information turns up, the story is revised, but the inadvertent bias favoring the first impression remains. When the narrative constructed in this manner is generally in tune with reality, then there is nothing wrong. If, however, the story is beside the truth, things can turn ugly.

These mechanisms, including the various pitfalls, take place in the head of every fireground commander who, shortly after arrival at the fireground, has to plot the strategy on the basis of the limited amount of information available at that moment.

4. FROM INFORMATION TO DECEPTION

Operational decision-making at a fireground is in essence nothing more than processing sensory cues on the basis of prior experience in order to produce a desired outcome. This mental process enables the fireground commander to size up the situation in a glance and decide upon the strategic deployment of the resources at hand or request additional ones if necessary.

When, for example, you are going to buy a new car, you have all the time in the world to gather the relevant information. Which make do you prefer and what model do you like? How many seats do you want? What is the gas mileage? Should you opt for the plug-in hybrid version? Or the electric version? And what about the options you would like to have? Leather or cloth seats? And so on. In principle, all the data you gather are reliable (although, with a car dealer and today's fraudulent software, you never know . . .).

In the best of possible worlds, we would have not all information but at least all relevant information at our disposal at a fireground. Why not all but just the relevant information? An overabundance of information makes it harder for us to know what precisely we should be looking at. An overload of detail constricts our thinking and will actually result in less understanding about what is really going on. More is sometimes better, but not always. Unfortunately, incident commanders rarely have all relevant information at their disposal at an incident ground, certainly not at the start of the incident. The reliability of the data at hand is another thorny issue. Furthermore, you have limited time to decide. A fast-moving and busy fireground is not the place to engage in lengthy cognitive debates. You have to deal with a lot of things on the fly and decide on the spot. Incident reports show that all the information the fire service did not have at their disposal at the beginning of the incident was by far more relevant than all that they did know or thought they knew.

In order of importance, the information that is unknown to us fire-ground commanders stands firmly on position number one. Translated to the fire of Beringen, that includes the following points:

- The raging fire at the front of the building hides a construction fire in the roof.
- The construction of the flat roof: Wooden beams were laid on steel castellated beams. On those wooden beams lay a kind of chipboard that was topped with tar and gravel. During the renovation works of 2007, thick petroleum-based insulation boards were placed on top of the old roof. The additional insulation was finished off with yet another layer of tar and gravel. A suspended ceiling was installed inside the supermarket. This resulted in a roof of 830 m^2 and approximately 80 cm thick that was not only waterproof but also completely airtight.
- The strongest side of the roof is the upper side.
- The fire started at the worst possible place, namely where the wooden roof structure of the later extension (building 6) joins the concrete vaults of the commercial block with flats at the front (building 3).
- The state of the building: Good craftsmanship was totally absent. The building had been shoddily renovated. It is a small miracle that long before the fire no serious problems with the stability of the building had arisen.

In terms of importance, it is a neck-and-neck race between unknown information and wrong information. In itself, there is no such thing as wrong information. Wrong information is nothing more than information or facts that are misinterpreted or from which erroneous conclusions are drawn by the incident commander or a third party who has provided that information. However, when decisions are based on wrong information, they tend to be erroneous.

The process of natural evolution has equipped the human mind with the magnificent ability to make rapid decisions in difficult circumstances based on an extremely limited amount of information that might even be irrelevant or misleading, namely the automatic mental processes of System 1. When we look at a series of facts or events we cannot help but weave a plausible explanation into them or impose a logical sequence onto them. Detecting patterns where none exists even has a name: *apophenia*.

We cannot deal with the whole concept of unpredictability. It is one of those mental flaws inherent to all of us. We find it hard to accept that things happen for no other reason than random chance. How ironic that mankind, which after all is little more than the result of an age-long process of natural selection and an improbable accumulation of coincidences, just cannot seem to come to terms with blind chance!

For this reason, the most idiotic conspiracy theories sell worldwide like hotcakes. In the Netherlands, conspiracy theorists deliberately set fire to 33 phone masts in 2020. According to them, the outbreak of the coronavirus was not the consequence of the unhygienic practices in Wuhan's Huanan seafood and wildlife market but was due to the rollout of 5G mobile phone services in the Chinese province of Hubei, of which Wuhan is the capital. In 2020, however, 5G had not yet been rolled out in the Netherlands.

Imposing logical connections on a sequence of events based on our own idea of how the world functions is an internal trait we human beings share with the other members of the family of the hominids. Chimpanzees, bonobos, gorillas, orangutans, and gibbons also convert sensory input into mental knowledge of the environment and are able to use that information flexibly. N. N. Taleb coined the term *narrative fallacy* for this tendency to weave a logical or coherent explanation into each sequence of facts we encounter.

"Situational awareness" has been for quite some time the buzzword at fire academies and in emergency services. But it has a flipside that is left in the dark. Situational awareness and narrative fallacy are in my opinion two sides of the same coin. At a fireground, luck is one of the many factors outside our control that play a role in the outcome of an incident. When the fire service does a fine job and everything runs smoothly, then our situational awareness was spot on. But if the incident ends dramatically, then narrative fallacy seems to be in play. Not uncommonly, we tend to disregard the role played by chance in our successes.

Research into the decision model rolled out by the National Fire Policy in the United Kingdom (situation assessment, SA; plan formulation, PF; and plan execution, PE) suggests that the three levels of situational awareness as defined by R. B. Gasaway (level 1: perception phase; level 2: comprehension phase; and level 3: projection phase) are rather theoretical concepts that are (too) far from what incident commanders of the fire service actually do on scene.[8]

Whereas Gasaway interviewed incident commanders after the incident had been taken care of, researchers for this British study filmed 33 incidents in real time. On scene, after having assessed the situation (SA), incident commanders immediately proceeded to implement their plan (PE) without explicit formulation of the plan (PF). And the phase of plan formulation (PF) turned out to be followed just as often by plan execution (PE) as by a reassessment of the situation (SA).

There is a big difference between what we do on scene and the theoretical concepts of decision-making that are taught at fire academies. The sooner the fire service can close the gap between practice and theory, the better. Do not get me wrong: situational awareness remains important. However, I am of the opinion that in decision-making at an incident ground, situational influencing of the adaptive unconscious by cues from the environment, the sense of the internal physiological condition of the body or gut feeling, and our own professional experience is much bigger and more consequential than we realize.

Our brain just cannot stop inducing and assuming various things automatically by means of analogies. In 1940, the political scientist E. L. Fordsworthe stated aptly that "the assumption is the mother of all mistakes." It should be noted, however, that although mistakes or errors are indeed almost always based on suppositions, not every assumption results in a mistake or error. Checking every supposition is simply not feasible in the everyday world, let alone at a fireground.

The following are some assumptions I made that night that turned out to be wrong:

- Having driven by the supermarket only 3 minutes before the emergency call was paged, I assumed the fire had started only a few moments ago. It turned out that the moment I passed the building on my way home, the fire had already been burning for more than half an hour. A CCTV camera in the neighborhood picked up the first signs of fire at 1:38 a.m. When dealing with a fire in an abandoned building, a fireground commander should assume that a longer period of time has elapsed between the start of the fire and the notification to the fire service because of the lack of habitation and social control in the building.

- An empty building implies that the fire load is limited. Indeed, this is true when you are thinking in terms of the incident type "fire in a building" because then you make abstraction of the building itself as a possible fire load. But with the incident type "building on fire" you come to the exact opposite conclusion. In Beringen, the roof and the interior finishing, for which excessive use had been made of sandwich panels and chipboards, created a fire load far above average.
- There was a residential block (building 5) in the retail estate on the side of the parking lot. I assumed that it would not be possible to enter the store through that building. During the renovation works of 2007, however, the load-bearing walls of that residential block were opened up and the ground floor was integrated into the supermarket. From the outside, there were no indications whatsoever that it was possible to enter the store through that residential block.

5. TO RECOGNIZE IS TO DECIDE

At a fireground, how do you as an incident commander plot the strategy under time pressure based on the limited amount of information available? Setting the strategy is like a job interview or a first date: there is no second chance to a first impression!

A lot of information enters the brain through the sense organs. Familiar cues are recognized as several competing neural networks inside your brain start to fire. Ultimately, one of these networks will gain the upper hand. Its particular pattern of neural activity will form your perception of the situation and shape your subjective experience. Operational decision-making in situ is essentially just a highly specialized form of pattern recognition.

In a split second you have unconsciously considered different strategies (e.g., offensive/defensive and interior/exterior attack), but this cognitive process ends as soon as you have found a strategy or course of action that brings about the goal you have set. You establish the strategy according to the principle of "to recognize is to decide" or *recognition-primed decision*, as demonstrated by the American psychologist G. Klein. I will return to the concept of priming in section 4 of chapter 3.

About 10 years ago the *predictive coding theory* was formulated in neurology and cognitive psychology. In this theory the brain is one big predictive organ. Just as the stomach digests food or the heart pumps blood through the body, the brain's job is to generate and to check predictions. These predictions, however, are not about the future but about the present: what is happening in the outside world that explains the input that is now pouring in. Our brain manages an internal model or *simulacrum* of the reality around us that it is constantly updating by checking sensory and other input against the situation we are in.

Our brain finds itself in the middle of a love triangle. It controls all our thoughts and actions from within the triangle:

- Environment
- Internal physiological condition of the body or *interoception*
- Prior experience

All day long our mind is continuously integrating situational cues, the internal physiological condition of the body, and prior experiences into a coherent account of what we are experiencing right now. An excellent and accessible introduction to the most exciting and promising theory of predictive coding is the fascinating book *How Emotions Are Made* by the American psychologist and neuroscientist L. Feldman Barret.

Unlike computers, our mind runs entirely on conflicts between these different predictions or best guesses that are constantly fighting for influence and competing for our conscious attention. There are always numerous possibilities, but ultimately one specific pattern of neural activity in your brain will win the argument and plot the strategy. In all likelihood the strategy that you have used in the recent past in similar incidents will prevail once more. The more successful that well-worn strategy has proved to be in the past, the greater your subconscious preference for it because, in human memory, recall is affected by recent experience and frequency. This effect is known as the *availability heuristic* or the *availability bias*.

However, the strategy of your choice is not necessarily the best of all conceivable strategies. It is just an effective strategy of which your memories are stronger and more accessible at that moment of time. Among the strategies that you did not consider, there might well be better ones, but under time pressure you did not even bother to take them into consideration or to conjure up a better alternative solution. When, after mental simulation, you have come to the conclusion that a certain course of action will work in the situation at hand, you stop reasoning through the wide array of options and implement the strategy your choice has fallen on.

In contrast to the unconscious serial evaluation of options, the just-mentioned mental simulation is a conscious thought process. Or, in Kahneman's terminology, considering the different strategies is an automatic or unconscious operation of System 1. The mental simulation or imaging how your strategy might be carried out, is a conscious mental process of System 2.

So "recognizing" makes an incident commander decide. The way it works can be illustrated with a slide loader. I know. That is outdated technology. Unconsciously and unwillingly my age starts to shimmer through. The idea is that of all the incidents you have attended to, your brain has made slides and those slides are stored in a mental slide loader. Together they form the unconscious mental background against which you consider the handling of the new incident. In the higher, associative areas of the cerebral cortex, previously acquired knowledge (intuition and automatic thinking patterns) and past experiences are stored. A certain cluster of nerve cells will send tiny electrochemical signals to another cluster with a certain pattern, during a certain time and in a certain frequency. This lets you unconsciously recognize the situation you are dealing with and allows you to instantaneously decide upon the strategy you will deploy.

I find this hypothesis convincing. I even have a fairly good idea of which slide in my mental slide loader might have lit up unwittingly that night in

August: the fire in the shop of the Total gas station on Wednesday, January 21, 2019.

In the late morning we respond to an emergency call. I leave the fire station in the command vehicle. Technicians had been connecting the wires of recently installed solar panels to the shop's control cabinet when it went awry. An energy arc passed through the control cabinet and set it ablaze. When I arrive, black smoke billows out the doors at the front and the back of the shop.

An interior attack is launched with bundles through the open door at the back side. The ladder truck is deployed so that we can check the condition of the flat roof. There are no flames shooting through the roof, but black smoke rises through various cracks and ventilation holes.

The attack crew inside the shop radios me that they have extinguished the fire. They request to start up the ventilation. On the outside, however, I notice that the black smoke keeps on billowing out the shop. There must be an additional seat of fire. I order the attack crew to keep searching for seats of the fire.

In the next room they find the burning control cabinet and extinguish it. The impact of their action is immediately visible. The color of the smoke becomes much lighter and the speed of the outflow decreases sharply. After a thorough check inside the shop, suspended ceiling included, the ventilation is finally started up.

Parallels with the fateful incident later that year on Coal Mine Lane abound: the choice for an offensive interior attack, setting up the ladder truck to check out the roof, the information from inside that does not match the image from the outside, and so on. There are also differences. At the gas station there was a rapid flow of black smoke, and the shop, with its surface of 220 m^2, was roughly four times smaller than the store expansion on Coal Mine Lane (building 6).

Thus an incident commander selects the strategy to apply, on the basis of the principle "to recognize is to decide" or recognition-primed decision. This perception-based decision process is, as I have just pointed out, a mix of unconscious and conscious modes of thinking. It is actually a textbook example of Kahneman's WYSIATI principle: a lightning-fast, intuitive conclusion. Your subjective confidence in that judgment depends almost entirely on the logical consistency of the story that your associative brain has spun quickly and automatically from the information at hand at that given moment (while neither the amount nor the quality of that information counts for much).

And now it becomes risky, for inattentional blindness and confirmation bias come into play. After you have decided upon the strategy, you will begin to look for those bits of information that are consistent with the strategy of your choice. You will focus on those details and aspects that make sense to you and assign too much weight to them, however few those elements might be. You will not notice or question or will reason away crucial pieces of information that are in conflict with your strategy. And what about gaps or blind spots in that strategy? Missing bits of information are effortlessly replenished by your brain with a logical alternative. Based on prior knowledge and expectations, top-down cognitive mechanisms in your brain will fill the blind spots that bottom-up mental processes have left blank. A (too) strong expectation can even evoke an observation. In February 2006, U.S. Vice President D. Cheney shot his friend during a pheasant hunt; he mistook him for a pheasant.

Because of the way our predictive brain works, we can be led astray. It can make us completely misinterpret what is really going on. Then we end up drawing conclusions for which there is insufficient evidence. Or even worse, we make decisions that are all wrong.

An example, on the night of Wednesday, January 8, 2020, the Iranian army fires two surface-to-air missiles near Tehran's Imam Khomeini International Airport. Five days earlier, Iran's top general Qassem Soleimani had been assassinated by the U.S. army in a targeted drone strike in Baghdad, Iraq. Intent on vengeance, the Iran military fires on the eve of January 8 a volley of missiles at two bases hosting U.S. troops in Iraq. The Islamic Revolutionary Guard Corps is on high alert for a potential U.S. retaliation. Once again the Middle East is living on the razor's edge. Because of failing radio equipment at the crucial moment, the operator of the air defense battery cannot contact his superior officer. He has to decide immediately and all by himself. He shoots the unknown incoming object that he sees on the radar screen out of the sky. In doing so, he thinks he is preventing an American missile strike on the Iranian capital, but in the following days it becomes apparent that he has shot down a departing commercial airliner. All 176 persons onboard were killed in the incident.

Translated to the Venn diagram in figure 5 the left circle is the ill-fated flight 752 of Ukraine International Airlines or the facts. The circle on the right is the alleged U.S. missile attack in the operator's mind or his automatic mental process. The blips he spots on the radar screen are the wedge in the middle both circles share. Note, however, that the figure below is a representation of the facts after the incident. Once again we have the luxury of knowledge afterward. The moment the operator decided to launch the missiles, he did not have that luxury. For him there was only the right circle. He had no awareness of the left circle. Just the idea of it was literally unthinkable for him, as he held his assumption or expectation to be true.

Amid the mounting tensions in the beginning of January 2020, it is hardly surprising that the operator interpreted the radar blip as an American missile strike. This tragic incident aptly demonstrates how context can dominate decision making.[9] It would be interesting to know the operator's training and background. Most likely, he was a soldier of the Islamic Revolutionary Guard Corps, the elite corps of the Iranian army. Military personnel are trained for combat. An air traffic controller might have shown greater reluctance

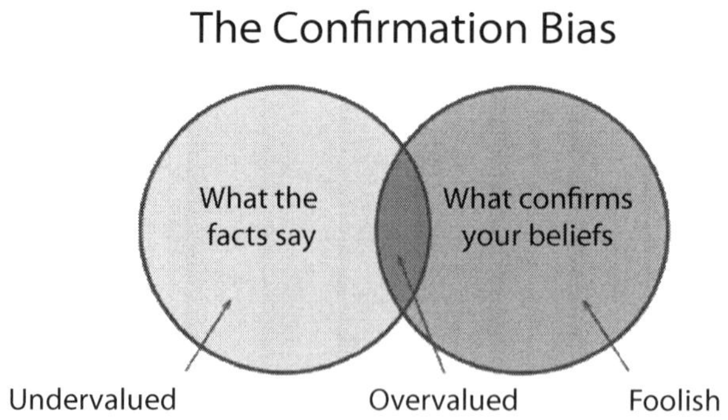

Figure 5. Confirmation bias

to deploy surface-to-air missiles because of his training and work experience. In his mind, the presence of Imam Khomeini International Airport in Iran's capital and the fact that an airplane transponder can also break down might have led to another outcome. Consider the case of Stanislav Petrov, the man who may have saved the world on September 26, 1983. Shortly after midnight, this Russian officer decided not to report computer readouts of a nuclear attack on the USSR by the U.S. to his superiors. He dismissed them as a false alarm: "He says he was the only officer in his team who had received a civilian education. 'My colleagues were all professional soldiers, they were taught to give and obey orders,' he told us. So, he believes, if somebody else had been on shift, the alarm would have been raised."[10]

What else can we deduce from the Venn diagram in figure 5? The size of the wedge in the middle is the decisive factor for a good outcome. The bigger the wedge at the moment when you make the decision, the more your perception/prediction/expectation is in line with what is really going on. The narrower the wedge, the more likely—the much more likely—you are misled and will end up making a flawed decision.

We can apply all of this—again with the benefit of hindsight—to the fateful fire in Beringen. The left circle represents all the factual information and the development of the fire as described in the incident report drawn up by the Federal Public Service Home Affairs. The circle on the right, on the other hand, is the fire development that I have put together based on the bits of information available that night and that resulted in the strategy that we applied. And what is in the wedge in the middle? The information I overvalued that night is the model BE-SAHF as it stands (the acronym BE-SAHF stands for building, environment, smoke, air, heat, and flame).

6. PHILOSOPHY OF SCIENCE FOR THE FIRE SERVICE

To get a grip on the world of great complexity we live in, we put everything in distinct categories, and by means of the scientific method we build models and hypotheses based on past experiences. Subsequently we try to verify and strengthen those models through systematic control.

And the fire service does exactly the same. In the fire academies in Belgium, for example, a distinction is made between three different kinds of fire development:

- Ventilated fire development
- Underventilated fire development
- Construction fires

Strategies for incident commanders and nozzle techniques for firefighters are linked to these different types of fire development. With such a static division into categories, the fire service risks losing sight of more than just the dynamics of fire. In itself the classification is not wrong, but along the way we have forgotten that categorizing always implies a reduction in complexity. Reducing the complexity of the world around us can entail considerable risks when sources of uncertainty are being ruled out. The result can even be that we no longer fully understand how the world really works because we are blinded by how little we know and grossly underestimate everything we do (and can) not know. This also applies to fires, as they are by far more complex than the simplified models we are taught at the fire academy.

One source of uncertainty at a fireground, for example, is that you as incident commander do not know for sure if your identification of the fire development is correct when you plot the strategy. And fire is furthermore an extremely dynamic process that simply cannot be locked up in one category. On the contrary, any fire will jump from one category to another when the conditions allow it.

That night in Beringen, long before my arrival on scene, the ventilated fire development in the abandoned supermarket was expanded with a construction fire in the ceiling. The raging fire at the front and the alleged kitchen fire inside the building concealed that construction fire. Each firefighter on scene was misled and we all got stuck in the wrong fire development.

Over the different kinds of fire development, the standard firefighting technique of reading the fire or the BE-SAHF model is applied in Belgium. By correctly interpreting these cues (building, environment, smoke, air, heat, and flame), the incident commander can assess the development stage of the fire and predict the future development. To comment in detail on all these cues goes beyond the scope of this book. I will limit myself to clarifying the element of smoke in a nutshell.

When you light the wood stove at home, you will have in the beginning light-colored smoke. In this phase, water vapor and decomposition gases are dominant in the smoke. In the early stage of a fire, the amount of smoke inside a room is limited. As the pressure builds up the light-grey smoke begins to escape slowly. Depending on the amount of fuel or oxygen available in the room, the smoke changes color during the development stage of the fire:

- In the case of a ventilated fire development, the smoke becomes darker and eventually will turn pitch black.
- In the case of an underventilated fire or a shortage of oxygen, the color of the smoke changes to yellowish brown.

As the temperature rises, the pressure builds up and the smoke starts to exit the room at an increasing speed.

The Austro-British philosopher of science K. Popper argued that a good theory can only be refined or refuted by demonstrating its falsity. He calls this *falsification* or the search for evidence to the contrary.

In our enthusiasm we have failed to falsify the BE-SAHF model. In all our live-fire trainings, we do little more than constantly verify the model within the same strict, well-defined settings. For example, the layout of the containers or the burn building of the fire academy used for live-fire training is meticulously described and defined for each of the various scenarios. The fire load is also carefully calculated because exactly the same amount of chipboards and wooden laths are used for each run. All these measures are necessary to ensure our safety during the training.

The BE-SAHF model is a theoretical model of reading the fire that has its origin in empirical observations made during fires attended by firefighters all over the world, but its field of application has never been established. As a result, that technique is now de facto applied to all fires in or of buildings as if it were a one-size-fits-all application. Were we to falsify the model, then the need to establish the field of application would immediately become apparent.

Thus, empirical observations are the starting point of the BE-SAHF model. However, if history has taught us anything, then it is for sure that learning from observations or experience has severe limitations and can lead to erroneous conclusions. Did not mankind think for a long time that the earth was a flat disk? And that the sun revolved around the earth? Even in the year 2024 we still fall into the trap of drawing the wrong conclusions from empirical observations. For the fire service, the Grenfell Tower Fire in London on Wednesday night, June 14, 2017, shows this at a macro level.

The incident report indicates that the Grenfell Tower Fire started at 12:54 a.m. on the fourth floor in the kitchen of flat number 16. The fire service arrives on scene 5 minutes later. But at 1:09 a.m. the fire escapes from the kitchen and gets into the external cladding that envelops the residential tower block for insulation purposes. Five minutes later, the fire service starts their interior attack in the kitchen of flat number 16. So even before the start of the interior attack, the fire development had already fundamentally changed. The ventilated fire development in the kitchen expanded with a construction fire.

Do you spot the danger of categorizing the fire development that I have just pointed out? For the sake of clarity, I emphasize that our London colleagues did not and could not have known that at that time. They did not have the benefit of hindsight as you and I have now. This example shows that in the event of a fire, a strategy can be outdated even before it has been deployed. To make things worse, incident commanders are unable to switch quickly between strategies. Because of inattentional blindness or confirmation bias, they just stick to their plan of action. As long as a fire development does not jump from one category to another, our strategy will be good. However, when a fire development changes too radically, incident commanders tend to get stuck in their first strategy and are prone to making erroneous judgments.

Even more interesting is the fire development in the exterior cladding of the tower block. Grenfell Tower was not the first high-rise residential block where the exterior cladding caught fire. What was new, however, was that the behavior of the fire deviated from the standard empirical expectation that we as a fire service had in the event of a fire in the exterior cladding of a tower block. What behavior did we expect?

In the past, this kind of fire always spread rapidly up the outside of the building, and once the soaring flames had reached the top, they simply went out by themselves. Such was the case in Dubai at the Torch Building and the Address Downtown hotel. In the exterior cladding of the Lacrosse Building in Melbourne the fire development was exactly the same. Initially, the fire in the Grenfell Tower behaves in a similar way. The fire climbs at a relatively slow speed of 4 meters per minute to the top. At 1:27 a.m. it reaches the upper floor. In 18 minutes' time, the fire has climbed from the 4th to the 23rd floor, and then something occurs that has never happened before.

Whereas in other residential tower blocks the fire stopped once it had reached the top, it now finds additional fuel in the architectural crown atop the building. Like the exterior cladding, that cornice was made of polyethylene. On exposure to heat, polyethylene melts and begins to drip. This petroleum derivative can, furthermore, flow while burning and generate burning droplets. At the architectural crown the fire splits in two directions. Gravity pulls the burning droplets down, and most of them end up in the insulation surrounding the tower. To make a long story short, the fire travels around the top; in the exterior cladding the fire does the same, but across the facades and from top to bottom. Within a few hours the fire engulfs almost the entire building.

Here I come again with K. Popper's falsification theory. Verifications of theories prove nothing; only falsifications can refute them. From all the fires in the exterior cladding of high-rise residential blocks before Grenfell Tower, the fire service should never have assumed that this would be the standard fire development.

Instead of assuming that fires in the exterior cladding always stop after having reached the top (and do not enter the building through the windows on their way up!), the fire service should have wondered why on the top of the building all those fires went out by themselves. It would have been fairly easy to figure out that in the past the fires stopped because of . . . lack of fuel. In other words, after falsification, the standard fire behavior that we assumed for all high-rise buildings before the tragic Grenfell Tower Fire can now only be upheld for those high-rise residential blocks where the roof is made of noncombustible material. In August 2021 this was confirmed in the southern outskirts of Milan, Italy. In the Torre dei Moro building, the fire could spread from the thermal covering on one facade of this 20-story apartment block to the panels attached to the other facade via the technical installations on the roof as they were made of combustible materials. By means of falsification, we define a field of application where this or that fire development is applicable.

Within the fire service there is more to falsify than just the fire development in the facade cladding of tower blocks. I already stated that the BE-SAHF model is based on empirical observations or experiences at firegrounds. Like each incident commander, I used that model that night in Beringen. And all my observations matched seamlessly with a ventilated fire development.

Upon arrival I see a raging fire at the front side of the abandoned supermarket. I conduct my 360-degree initial size-up around the structure. At various places I perceive grey-black smoke slowly leaving the building. That does not surprise me. The fire at the front causes the air pressure to rise inside the supermarket, and that overpressure slowly pushes the smoke out through various openings. After the fire at the front is extinguished, the smoke keeps coming. There must be a second seat of fire. The team that I in the meantime have sent in via Saint Barbara Street reports a kitchen fire. This confirms my idea of a second seat of fire. Everything fits perfectly in the BE-SAHF model because I can always link the smoke to a seat of fire. But then the scenario changes and from one moment to the next nothing makes sense anymore.

Where is the flaw in the model for reading the fire? In my opinion we have overlooked that each time an incident commander reads the fire, he is not aware of the fact that he sees the incident through a particular pair of glasses. And those glasses are the *incident type* that he has determined subconsciously based on an intuitive WYSIATI decision. I propose to add an "I" to the model of reading the fire. Hence I speak of I-BE-SAHF, where the "I" stands for two things: intuition and incident type.

What exactly do I mean when I say that the incident type intuitively or unconsciously colors—or even dominates—your reading of the fire? To clarify this, I will apply it to the type of building I had to deal with that night in Beringen: a large building. In my opinion, you can roughly choose from four types of incidents:

- Type 1: Small fire(s) in a large building
- Type 2: Construction fire in a large building
- Type 3: Great fire in a large building
- Type 4: Large building on fire

Type 1: When you are dealing with a small fire in a large building, the fire (load) is limited: for example, a machine on fire in a factory building. For such a small and manageable fire in a large building, the standard deployment is an offensive interior attack. I call this type *a fire in a building.*

Type 2: A construction fire in a large building implies that the construction of the building itself is the fire load. And that fire load can be huge. If you cannot get the situation under control with piercing nozzles or Fognails in an early stage of the fire development, a defensive exterior attack might well be the only safe strategy. This type I call *a building on fire.*

Type 3: In the event of a great fire in a large building, the fire (load) in the building is huge: for example, a fully developed fire in a mega furniture store. More than likely the fire will not be restricted to just the contents of the building because the building itself will also catch fire. Here the fire service has to opt for a defensive exterior attack.

Type 4: In a large building on fire, content and construction together form the fire load. An example of this type of fire known to every Belgian firefighter is the tragic fire of the Innovation department store in Brussels in 1967. A defensive exterior attack is our only option.

A large building on fire (type 4) and a great fire in a large building (type 3) are easy to recognize. The distinction between a small fire (or fires) in a large building and a construction fire in a large building is less straightforward, particularly when that large building is an empty building, as was the case in Beringen. The cues emitted by a fire in a building and a building on fire may be identical, but these fires require entirely different approaches.

The subconscious choice of the incident type by the incident commander based on these cues is an automatic mental process of System 1. Prior experiences and training are the personal framework of each incident commander. I think that this intuitive choice has already been made long before the incident commander has conducted an initial size-up at the fireground.

When you spot from afar an immense black plume of smoke on your way to a factory fire, for example, you are already scaling up before you have arrived at the incident ground. At the latest a few moments after your arrival on scene, but surely before the start of the 360-degree initial size-up around the building, you unconsciously get a first impression of the situation and assess the incident.

Under the radar of consciousness and by means of *selective information reduction*, your unconscious has effortlessly and rapidly reduced the observed situation to thin slices of information and filtered out the essential cues. In this manner an incident commander intuitively grasps the severity of the fire and assesses whether they can or cannot attend to it with the material at hand and with this strategy or the next. Then the incident commander does a 360-degree size-up and starts to look for additional information that is consistent with their first assessment of the situation or that impels them to adjust the strategy. Compare, for example, how in speed dating or casual encounters on the street, your first impression of the other person is already formed based on perceived facial expressions and attitude long before you have talked to that person. But there is catch, as the social and cognitive psychologist J. Bargh has noted:

> It's a person's behaviour that matters, not their face. We all know this at an intellectual level, of course, but is very hard to shake the impression we get from a person's face, especially our first impression. It is not so much that we think we know what a person is like based just on their face. It is that we feel absolutely certain that we are right about what we think about them.[11]

A similar catch applies to a fireground: It is the *incident type* that matters, not the *visible cues of fire and smoke* assessed in a glance upon arrival. First impressions can be deceiving.

This cognitive process is clearly discernible in the incident report of the fire at De Punt in Drenthe, the Netherlands (2008):

- The fire truck leaves the fire station at 2:14 p.m. on May 9 in response to a house fire. However, it concerns a fire in an indoor storeroom with a depth of 15 m that is located in a large hangar (25 m × 75 m) in which recreational boats are maintained and stored.
- Two minutes later, when the fire truck is approaching a roundabout at the rear of the hangar, the crew sees a thick layer of brownish-yellow smoke quietly flowing out the rear of the hangar.
- The warrant officer in the fire truck scales up to "medium fire." An additional unit and the officer on duty are dispatched to the fireground.
- The fire truck arrives at the scene around 2:18 p.m. In front of the hangar the situation is completely different. There is hardly any smoke; just a few patches of smoke are visible at the back of the hangar.
- A police officer informs the firefighters that there is a car burning in the back of the hangar. The warrant officer assesses the situation again.
- The house fire, which only a few minutes ago was intuitively scaled up to a "medium fire," is now downsized to a car fire by exactly the same automatic mental process of System 1.
- After the warrant officer has given the formal order "explore and extinguish any fires," four firefighters enter the hangar and the warrant officer starts his 360-degree size-up around the building.
- At 2:21 p.m. the warrant officer scales up to "big fire." The entire hangar is ablaze and three colleagues are still inside the building. The three firefighters perish.

Typical of this incident are the big differences between the cues picked up by the warrant officer when he and his colleagues were driving up to the fire and those picked up upon arrival at the fireground. Time and time again, he intuitively assessed the dynamically changing situation all over again, and this only on the basis of the limited new information that had become available. His strategies form, as it were, a series of consecutive but completely distinct WYSIATIs. The scaling up during the approach is no longer relevant when they have pulled up in front of the hangar and the incident is intuitively reduced to a car fire.

This sequence of events is fully in line with Kahneman's theory of the automatic operations of System 1. Central in the process of decision-making stands the coherence of the story that the warrant officer has managed to put together, but that narrative is based on whatever information happened to be available. The amount and quality of the data on which that interpretation is based are largely irrelevant. The previous intuitive impressions/judgments could not play a significant role as only a few seconds later they were put into question and subsequently discarded based on completely different sensory cues. That day, the intuitive judgments, which neglect ambiguities and suppress doubts, undeniably worked to the disadvantage of the fire service.

The theory taught in fire academies is that you do the 360-degree size-up before establishing the strategy. However, the practice is completely different. Long before an incident commander does their size-up around the building, they already have a first impression of the incident and a broadly outlined course of action. Because of their unconscious tendency toward self-confirmation (confirmation bias), those elements that confirm their first assessment will easily capture their attention during the size-up. Signals that are in conflict with that assessment are more prone to be overlooked, disregarded, or explained away. And the longer the incident commander has used a theory or strategy, and the richer and more consistent their narrative becomes over time by all the details and the new information they have managed to integrate, the harder it gets to puncture that precarious bubble of improbabilities. The problem with theories and strategies is that they are inherently sticky. Once you have put together a theory or strategy and used it as a tool in your thinking, it is virtually impossible to notice its inherent flaws. Kahneman speaks of a *theory-induced blindness*. Even psychiatrists who are well versed in the working of the mind are easy to catch out.

For a mischievous experiment, psychologist D. L. Rosenhan and seven other sane pseudopatients (including two other psychologists, one pediatrician, and one psychiatrist) were to gain secret admission to 12 different psychiatric institutions. In the hospital, all eight participants had to cease simulating symptoms of mental illness and answer the questions honestly. Only once were they allowed to tell a lie: When registering at the admission office, they were instructed to say they had been hearing voices. Most of the time the voices were unclear, but "empty," "hollow," and "thud" were some of the words as far as they could tell. The results of this experiment, published in 1973, are telling.

- Upon admission, all but one of the pseudopatients were diagnosed with schizophrenia.
- Getting discharged from the hospital turned out to be a lot harder than getting admitted. On average, they were kept 19 days in the psychiatric facility. One pseudopatient was discharged after no less than 52 days!
- At discharge, the diagnosis for all pseudopatients was "schizophrenia - in remission," which means that the symptoms are under control but the patient is still suffering from schizophrenia.
- The pseudopatients were never detected by the psychiatrists or the nurses. Only the other patients in the hospital—the real psychiatric patients—smelled a rat and voiced their suspicions. They regularly said to the pseudopatients: "You are not crazy. You are a journalist, or a professor. You are checking up on the hospital."

For the second part of this study, Rosenhan made arrangements with a research and teaching hospital whose staff doubted that such large errors could occur in their hospital. Again pseudopatients would be sent, and the staff members had to identify them. Over the following 3 months, the hospital staff identified as many as 41 of their 193 new patients as possible pseudopatients. With respect to 19 patients they were fairly sure. Rosenhan, however, had not sent one single pseudopatient.

How the incident commander interprets the cues of the fire upon arrival at the fireground is a direct and powerful determinant of the strategy that will be deployed. That night in Beringen, everyone thought we were dealing with small fires in a large building. All the signals emitted by the fire (e.g., smoke, flames, airway) were read and interpreted as though this were the scenario, and this led to interior firefighting. If upon arrival the incident type had been recognized as a construction fire in a large building, the same cues would have been read completely differently and we would have handled the incident in another manner. The intuitive choice of the incident type—the "I" of I-BE-SAHF—determines how the fire should be read and how it effectively will be read.

Although further research is required, I have the impression that incident commanders assess the type of incident by intuitive thinking or another unconscious mode of thinking of System 1 that is based on their first impression upon arrival at the fireground. This way of working is not necessarily bad. Or, in the words of Kahneman himself,

> System 1 is indeed the origin of much that we do wrong, but it is also the origin of most of what we do right—which is most of what we do. Our thoughts and actions are routinely guided by System 1 and generally on the mark.[12]

This identifies the basis of the strategy the fire service deploys at an incident as the result of an intuitive mode of thinking. But when can you as an incident commander trust your intuitive judgment? That answer is to be found in the basic conditions for acquiring a skill:

- An environment that is sufficiently regular to be predictable
- An adequate opportunity to learn those regularities through prolonged practice
- Rapid and unequivocal feedback about the correctness of your thoughts and actions

When these criteria are met, your intuitions are likely to be skilled and mostly valid. But are all these criteria currently met in the landscape of the Belgian fire service? The reform of the Belgian fire service in 2015 has significantly augmented not only the number of hours that we train but also the quality of our training. And that is good, very good indeed.

As long as the cultural milieu in which the fire service has to operate does not change too radically, incident commanders will on average tend to make the right decisions. But is our cultural milieu still sufficiently predictable? Has that not changed too rapidly and too drastically?

Nowadays, society and the risks associated with modern-day life change at an ever-accelerating pace. Vehicles on alternative fuels are already driving on our roads long before our firefighters are trained to deal with these new risks. Ditto for new construction methods. To improve the energy performance of our dwellings, we make them more airtight and insulate them better than ever before. This inevitably results in other fire developments. Not only the number of underventilated fires but also that of construction fires will increase over the years to come. You see that trend already today. The number of homeowners who have accidentally set fire to the insulation in the cavity wall of their own house while burning weeds can no longer be counted on one hand. A thorough analysis of the fire development of construction fires is necessary, as well as extensive special training on it.

7. IS THINKING DOING, OR DOING THINKING?

At the Belgian fire service, human resources training is for the time being a too small part of the training program. Situational leadership, the different hats of the manager, how to conduct performance evaluations, and so on are just a few of the topics that are being addressed. The training offered falls roughly under the topic of providing leadership to a team in a fire station.

You will find little on the biophysical or psychological mechanisms that come into play when you, as an incident commander, plot the strategy at an incident. Even less is said about how emotions—or better, innate physiological reactions—completely take over your thinking and acting when, because of an unexpected turn of events, you suddenly no longer have control over the incident at the fireground. It is high time to examine how our brain operates under those extreme conditions.

Investigations into incidents within the fire service have repeatedly established that the incident commander is unable to switch between strategies at crucial moments during the course of the incident. The incident report of the Grenfell Tower Fire, published in 2019, states unequivocally: "The first incident commanders, although experienced, were of relatively junior rank. . . . None of them seem to have been able to conceive of the possibility of a general failure of compartmentation or of a need for mass evacuation; they neither truly seized control of the situation nor were able to change strategy."[13]

I find this judgment extremely harsh. A total of 272 residents and visitors managed to leave the burning residential tower block that Wednesday night. Unfortunately, 72 persons perished. The difficult work of our London colleagues seems to be judged exclusively on the outcome of this tragic fire rather than on beliefs that were reasonable at the time the London Fire Brigade was dealing with the fire in the Grenfell Tower. Two questions arise:

- Is this the task of the first incident commanders of relatively junior rank?
- Is it realistic to assume that the stay-put policy could have been revoked earlier that night?

Translated to the ranks used in the Belgian fire service, the first incident commanders were warrant officers with the rank of adjutant. These incident commanders were first engaged in the kitchen fire in a flat on the fourth floor, and then additionally they had to deal with the fire in the exterior cladding. Organizing a mass evacuation and thus revoking the stay-put policy is in my opinion a task for the incident commanders of higher rank that were later called to the scene and took over command.

Irrespective of the question of who should have made that judgment, it is more interesting to question whether the first incident commanders or even the senior fireground commanders who arrived on scene later could have made the decision to mass evacuate sooner. Proceeding to mass evacuation is, after all, completely at odds with the strategy that the London Fire Brigade applies to fires in residential tower blocks. Their advice is to stay put and let the compartmentation of the building do what it is designed for. In a residential tower block, each flat forms a compartment that is designed to withstand fire for a certain period of time. Staying in your flat with the door closed and waiting for the fire service to do its job is the basic principle of the stay-put policy.

Before the Grenfell Tower Fire (June 2017), a coroner's inquest made recommendations in the aftermath of the Lakanal House Fire (July 2009) to review the stay-put policy and consider a mass evacuation when necessary. However, this led to insufficient changes in their standard operating procedure for a fire in a high-rise residential block—not only at the fireground but also in the call management by the dispatchers, and in their fire survival guidance calls in particular. It is therefore no coincidence that the decision to mass evacuate that night was made only 1 hour and 53 minutes after the start of the fire.

Neurological research indicates that under stress and time pressure it is very hard for the human brain to switch from one strategy to another, particularly when, for example, the desired output is the exact opposite of the standard operating procedure that has been drilled for years and used time and time again. Stress affects the ability of each one of us to think rationally and make good judgments.[14] Under strain, you regress to what you have learned first and know best. For our London colleagues that was the policy for the residents to stay put.

Setting the initial strategy is therefore extremely important. If that is correct from the onset—that is, your interpretation of the events is in line with the actual situation—then everything will turn out fine. If you take a wrong turn from the very start, you will more than likely get stuck in your flawed strategy. The decision to switch is anything but obvious, if it comes at all. Most likely that switch will come from a fellow officer that is called later to the scene to offer assistance rather than from the first incident commander. This was the case with the Grenfell Tower Fire.

- After having seen footage of the burning tower on Sky News, dispatch advises callers from inside the high-rise residential block no longer to stay in their flat. Their advice or fire survival guidance is, from 2:33 a.m. onward, to leave the building if possible.

- Ten minutes later, a high-ranking fireground commander who had just arrived on scene and taken over command does exactly the same. He revokes the stay-put policy.

The dispatcher in charge and the high-ranking fireground commander that arrived 90 minutes later on scene were able to do something that the first incident commander no longer could do. Both could make an unbiased, new assessment of the situation at hand with distinctly less stress in their body. As I have said before, setting the strategy at a fireground is like a job interview: there is no second chance to a first impression!

I do not think it is a coincidence that comments on the lack of switching between strategies abound in incident reports. The moment an incident commander has lost control over the incident, you must not expect them to calmly and rationally reassess the situation. In that state of mind they simply cannot come up with a new operational strategy.

What happens in the head of a person when they lose control over the situation from one moment to the next? In any case, they are out of their comfort zone, feels uneasy, and might even become anxious.

Research indicates that your state of mind is a crucial factor for having a sudden flash of insight or coming up with creative, novel solutions that might solve the problem.[15] When you are under stress, your brain can only concentrate on details and you fail to see the broader picture. On the other hand, when you are in a good mood, your brain—in this case the prefrontal cortex—is better at finding creative solutions to the problems you are facing.

The prefrontal cortex is the outermost layer of the frontal lobes. This brain area is the last to fully mature, and it does that around the age of 25. Parents of teenagers experience this every day. It is precisely because of this late development that your lovely daughter or reckless son does not mince words during puberty, knows everything better, and carries out everything that comes to mind. The prefrontal cortex allows you to rationally solve problems, think ahead, make decisions, behave socially, and repress impulses.

These functions have all been known since September 13, 1848. That day, railroad foreman Phineas Gage got a metal rod over a meter long right through that brain area. With that rod, this mild-mannered and hardworking model worker had been tamping a blasting charge into a borehole. In doing so, he must have caused the spark that set off the explosion. Miraculously, he survived the incident and did not suffer from paralysis, speech disorder, or amnesia, but Phineas was no longer Phineas. After the incident, he had a completely different personality. John Harlow, the physician who treated him and reported extensively on his case, put it this way: "He was fitful, irreverent, indulging at times in the grossest profanity which was not previously his custom, manifesting but little deference for his fellows, impatient of restraint or advice when it conflicts with his desires, at times pertinaciously obstinate, yet capricious and vacillating, devising many plans of future operation, which are no sooner arranged than they are abandoned. . . . A child in his intellectual capacity and manifestations, he has the animal passions of a strong man."[16]

In 1962, the brain of a soldier was damaged in exactly the same spot. During a mission abroad, he drove his jeep over a landmine and a metal edge from the windshield penetrated his forehead. He too lost everything because of his impulsive, socially unacceptable, and emotionally clumsy behavior. It is not surprising that the lobotomy, a surgical procedure during which the frontal lobe was cut loose from the rest of the brain, quietly fell into disgrace around the middle of the 20th century. More than once an unmanageable psychiatric patient or occasionally a pubescent teenager was turned into an apathetic greenhouse plant with no interest whatsoever in anything around them.

When processing sensory information, the brain uses a two-track policy. In the center stands the thalamus in the limbic system, which functions as the gateway to the cerebral cortex. All information that enters the body through the eyes, ears, or skin travels in spikes to the thalamus. The functioning of this brain region can be compared with a switching station. It filters incoming information and passes it on. Under normal circumstances, the thalamus transmits the information to the cerebral cortex, where it is further analyzed and enriched. A check file goes to the amygdala for confirmation that the followed pathway is the correct one.

The amygdala is the "threat center" of the brain. All the time it is vigilantly monitoring our environment for possible dangers, and in less than two hundredths of a second it assesses the emotional charge of each image (danger/no danger). That speed is several times faster than conscious perception, and even too fast for a person to have conscious knowledge of the image. Thus we enter the field of subliminal perception or perception that falls below the perception threshold of consciousness. This kind of perception is exquisitely sensitive to negative information. The information from the amygdala feeds directly into the part of the brain that processes emotions, bypassing the visual cortex that supports our conscious experience of seeing. This is a good example of an unconsciously perceived external stimulus that is processed top-down. If the emotional charge is positive, then you do what you think you must do. I summarize this trajectory as "first thought and then done." In Kahneman's terminology, this is a conscious process or System 2.

In situations of risk, danger to life, or fear, the information flows exactly the other way around. Typical for this kind of situation is an increased stress level. Under stress, the thalamus sends the information first to the amygdala, which immediately initiates a response: fight, flight, or freeze. Then the information is passed to the cerebral cortex, where to the extent possible the initiated reaction will be reconsidered as appropriate or not. I summarize this alternative trajectory as "first done and then thought." It takes place entirely outside our conscious awareness and is one of the automatic operations of Kahneman's System 1.

What is the reason for this alternate route? A conscious, rational analysis when you are in a life-threatening situation is simply too slow. Speed is now of vital importance, and to engage in lengthy logical considerations is the last thing you must do.

Suppose a car is speeding in your direction. Should you jump to the right or to the left? Or is diving away better? By the time you have rationalized the various options and decided what to do, you will be overrun. To have any chance of survival, you have to react right away. This speed of working is typical not just of a primary emotion such as fear; the same applies to other emotions such as joy, sadness, anger, shame, and so on.

I use the term emotion in the sense defined by the American evolutionary psychologist R. M. Nesse:

> Emotions are specialized states that adjust physiology, cognition, subjective experience, facial expressions, and behavior in ways that increase the ability to meet the adaptive challenges of situations that have recurred over the evolutionary history of a species.[17]

You have undoubtedly already experienced yourself that in the initial stage the emotion of fear has a short-term inhibitory effect on your body. Your heart rate and breathing are suppressed for milliseconds. You prick up your ears and your pupils dilate. Then you get hot and start to sweat. Within a fraction of a second, pulse quickens, blood pressure rises, blood vessels dilate, and the airways in the lungs widen. Digestion stops, as now all the blood in your body has to flow primarily to the muscles. All these changes prepare the body for an emergency action: fight or flight.

Emotions are evoked at lightning speed, much faster than our reflective thought and even our conscious perception. They are already directing various behavioral systems long before you are consciously aware of it. You never witness the beginning of your emotion; you only experience the feeling or the reaction/behavior it causes. This gives the false impression that the whole (stimulus and reaction) is a conscious process. For firefighters at a fireground, the fear reflex is by far the most important innate physiological reaction.

8. THE MISMATCH BETWEEN BRAIN AND ENVIRONMENT

In their book *From Big Bang to Burnout* (*Van big bang tot burn-out*), the Dutch psychiatrist W. Hoogendijk and the journalist W. de Rek describe the development of the human stress hormone axis throughout evolution. Books on evolutionary theory are rarely best-sellers, unless you have succeeded in seamlessly linking our species' long evolution to contemporary stress-related mental health conditions like depression, hypertension, and burnout. The cause of these and various other mental problems is precisely our evolutionarily developed stress response system, which is simply not (yet) adapted to the continuous and incessant flow of stressors that typifies modern-day life.

Maybe over hundreds of generations or several tens of thousands of years, natural selection might have enough time to do its work and engineer a successful genetic adaptation. Whereas you used to study the past to understand the present, you do it now, thanks to Hoogendijk and de Rek, to cope with the overdose of stressors with which mankind has saddled itself. I strongly recommend this book.

Like all other reflexes, the fear reflex or the survival circuits in our brain were shaped and honed over millions of years of evolution and exquisitely refined by natural selection over time. For our current stress response system, everything began 500 million years ago with the appearance of a new species on planet Earth: the fish. Fish were the first animal species with nerve cells that specialize. Together they form some kind of primitive brain. Evolution by natural selection is very slow, but the cultural environment we live in can change rapidly. We have been living in a technical society with cities and overpopulation for only a few thousand years, but the evolution of humans and our brains expands over a period of 100 million years.

Today's reflexes or unconscious mechanisms were created in different times and places from those we currently live in. It all dates back to the time when ape-people like Lucy (*Australopithecus afarensis* or the southern ape of Afar) roamed around the African savanna some 3.2 million years ago. More than 100 parts of Lucy's skeleton were dug up in 1974 during excavations in Hadar, an archaeological site on a tributary of the Awash River in the Afar region of Ethiopia, Africa. This ape-person is named after the pop song "Lucy in the Sky with Diamonds," a track on the iconic album *Sgt. Pepper's Lonely Hearts Club Band* of The Beatles. Just for fun, the archaeological team ended the first working day by playing this psychedelic song loudly and repeatedly in the excavation camp the whole evening long. For all we know, we could have ended up referring to her as Lady Madonna or Eleanor Rigby.

Lucy and the other members of her species closely resembled apes. In some respects they were also more ape than human. Nevertheless, ancestor primates like Lucy distinguished themselves clearly from apes: They walked erect on two legs with a striding gait rather than moving in a knuckle-walking fashion like the gorilla. The volcanic ash near Laetoli, Tanzania, contains an about 3.6-million-year-old trail of footprints left by three hominins, probably *Australopithecus afarensis*.

The weight of Lucy's brain was exactly that of modern chimpanzees or orangutans: roughly 500 grams. Over the next 3 million years, the weight of our brains tripled to 1,400 grams and the number of nerve cells (*neurons*) increased to an estimated 86 billion, each having 10,000 *synapses* or connections. As a result, we now possess a brain that is about six times larger than it should be for a mammal of similar size. The size of our brain is appropriate for a tyrannosaurus. Ancestor primates and the great apes of today have brains about twice the size expected for an average mammal. On top of what was needed to direct a body, our already large primate brains swelled even more, allowing us as a species to communicate more accurately with each other and to master the challenges we faced in a world that was growing ever more socially complex at an infernal pace. By simply walking erect and expanding a strict vegetarian diet with eating occasionally meat, the *Australopithecus*—the species to which Lucy belongs—took the first step.

Walking on the hind legs freed up the hands for handling tools and countless other sophisticated labor activities. The development of hunting also played a major role in our transition from animal to human. The hunting revolution gave us more than just animal protein and fat rich in energy and essential fatty acids. For a successful hunt rudimentary weapons had to be fabricated, and a good relationship, cooperation, and communication between the hunters were also indispensable. In the camps of our hominin ancestors, reciprocal altruism gave rise to a division of labor.

While the hunters brought meat to the camp, and the fishers fished, others collected plant nutrition (roots and tubers) and the grandparents gathered wood for the fire with some help from the children. In short, in our distant past, there have been more than enough social challenges and sophisticated manipulations with tools to stimulate an unprecedented growth of the brain, which conceivably was thrown into a higher evolutionary gear by sexual selection.

The natural selection and biological evolution of reflexes and intuitions in our steadily expanding brain was driven by solving important recurring problems. The DNA or molecules of fish, land animals, mammals, apes, ancestor primates, and finally the first hominids with, for example, no fear whatsoever for large predators, poisonous snakes, dangerous spiders, or massive wildfires were weeded out of the gene pool generations ago. These somewhat overconfident ancestors did not survive their confrontations with those dangers as they did the wrong thing at the wrong time. That is natural selection at work. Depending on environment (savanna or riverbank), circumstances (day or night), and prior experiences (water extinguishes fire), our brains worked out sustainable solutions.

Around 200,000 BC the size of our brain became what it still is today. At that moment speech also seemed to make its appearance. The *Homo sapiens* set foot on African soil and from 80,000 to 50,000 BC immigrate in different waves to Europe and Asia. Our unparalleled ability of acclimatization and cultural adaptation to the environment enabled us to move on and to thrive in even the most inhospitable physical conditions on earth (e.g., the San in the Kalahari Desert or the Inuit in Greenland). Meanwhile on mainland Europe, some 7,700 years ago a few genes of these African hunter-gatherers with dark skin mutated. In the relatively dark Europe with its weaker sunlight it was difficult to produce vitamin D, so depigmentation took place and the color of our skin lightened.

The hasty leap of this unimportant new species to the top of the food chain is masterfully described by the Israeli historian Y. N. Harari in his book *Sapiens*. Wherever *Homo sapiens* gained a foothold, they wiped out most of the local megafauna and left behind a mass of archaeological artifacts. They were constantly looking for nodules of silex that are easy to process, and with targeted blows they turned them into convenient stone tools.

The evolution of modern man or the *Homo sapiens* took hundreds of thousands of years. In that long process, the sustainable solutions evolved into automatic modes of thinking that, once encoded in our genome, were passed down the generations. If similar threats occurred often enough, the emotion of fear would have been often—although not always—a good solution.

A natural reflex is and remains a matter of averages and probabilities. There is no 100% guarantee that it will be the right reaction in all circumstances, especially as time goes on. The survival reflex we fall back on today in our high-tech world when facing a threat to life is nothing more than the average solution to an average problem our prehistoric ancestors encountered regularly. It goes without saying that the risks we ran in our tribal, hunter-gatherer days are nothing at all like those in the urban jungles of today.

Now what is the main problem for a fireground commander during incidents? From an evolutionary point of view, acute stress is developed for a rapid response to a short exposure to danger and intended to gear your body up for immediate action. Heart rate and blood pressure must rise because this is the only way that the muscles in your body will get the necessary energy so they can tense up and enable you to flee or to fight. Controlling your blood pressure is precisely the main function of the hormones *cortisol* and *adrenaline* in this physiological stress response.

However, as soon as an incident commander has completely unexpectedly lost control over an incident and things start to go horribly wrong, this physiological reaction can spin out of control. The healthy stress you experience during every incident changes in no time into an unhealthily high stress level, fear, or even panic. Instead of a brief exposure, you are dealing with a prolonged state of disproportionally high stress. When the stress is too high or the exposure too long, more than just your heart rate and blood pressure go up. In your blood, the levels of adrenaline and the stress hormone cortisol begin to spike. That is not only unhealthy; it also has a huge impact on the working of your brain. Unquestionably you have already experienced yourself in daily life that in periods of acute distress or when you are anxious, you no longer can think calmly. What is the cause of this?

The stress hormone cortisol and the neurotransmitter noradrenaline seriously disrupt the functioning of your working memory or your cognitive performance. When you plot the level of cortisol in the blood on the X-axis against the learning capacity on the Y-axis on a graph, the curve is an inverted U (fig. 6). With too little arousal you learn nothing, but with too much you are paralyzed by stress. You need the right amount of this chemical.

A good level of cortisol or healthy stress stimulates the learning curve in the beginning until the optimal amount is reached in the middle of the inverted U. After that point, the increasing stress level begins to hinder learning and eventually will cripple it completely. Exactly at the moment

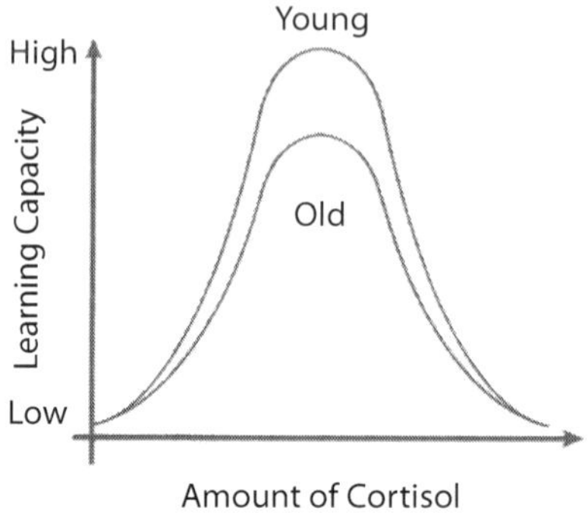

Figure 6. The relationship between the stress hormone cortisol in the blood and the learning capacity

when a life-threatening problem needs to be addressed urgently, all kinds of hormones and neurotransmitters suppress the conscious mental processes of System 2. The evolutionary reason for this bypass is saving energy. Keep in mind that in the body the brain is the major consumer of oxygen and glucose. This organ, which constitutes on average 2% of the body weight, consumes no less than 20% of the body's energy. In the event of danger, all the glucose available in your body has to be ferried to your muscles because now your priority is to flee or fight and to learn so you can react more appropriately when later on in life you are confronted with a similar problem. You get stuck in an excessive fear learning process and develop what is known as *perceptual narrowing*. Your attention is restricted to just those details that frighten you. All other things fall outside that focus and simply cannot be taken into account. All contextual information evaporates.

In this emotional state of mind you must not expect much rational thinking or logic. This does not mean that your brain breaks down completely. It still functions, but for the time being it does not work normally, and this prevents you from functioning properly. Moreover, you will not even notice that parts of your mental capacities are bypassed as everything feels and seems to be operating normally. The best description of one's state of mind at such a moment comes from M. Gladwell: You suffer from *mind-blindness*.

The list of the incidents in which your colleagues or yourself as an incident commander might experience such a mental breakdown can be drawn up easily. The trigger is always an unusual twist of events at the incident ground or the continued lack of an anticipated change, which results in a situation that is beyond your control. The more gruesome the circumstances you find yourself in and the longer you are exposed to those harsh conditions, the greater the impact of stress on your brain.

In addition to the classic cues of unpredictability, loss of control, length of time, and horror, firefighters also have to be aware of emotional involvement. Too much empathy for or emotional involvement with the victims of an accident will result in a psychological overload. Fatal accidents involving young children are strenuous for every firefighter. Even harder to handle are incidents where the victim turns out to be an acquaintance or a member of your own family or incidents where fellow firefighters suddenly become victims themselves because of an unexpected twist in the scenario.

Colleagues—from firefighter to officer—who succumb to the stress during incidents and become anxious or start to panic are lost for the remainder of the incident. They must always be replaced by a colleague to avoid further psychological damage from prolonged exposure to an unhealthily high stress level. More than likely they will leave the incident ground with an *acute stress disorder*. This is the short form of the better-known posttraumatic stress disorder (PTSD). Just like PTSD, an acute stress disorder is a mental health condition that can occur in people who have experienced or witnessed a traumatic event. According to the American Psychiatric Association, the symptoms of an acute stress disorder manifest between 3 days and 1 month after the event. The various symptoms fall into four categories: [18]

- **Intrusion**: Intrusive thoughts such as repeated, involuntary memories; distressing dreams; or flashbacks of the traumatic event. The flashbacks may be so vivid that people feel they are reliving the traumatic experience or seeing it before their eyes.
- **Alterations in cognition and mood**: Inability to remember important aspects of the traumatic event; negative thoughts and feelings leading to ongoing and distorted beliefs about oneself or others; distorted thoughts about the cause or consequences of the event leading to wrongly blaming oneself or others; ongoing fear, horror, anger, guilt, or shame; much less interest in activities previously enjoyed; feeling

detached or estranged from others; or being unable to experience positive emotions.

- **Avoidance**: Avoiding reminders of the traumatic event may include avoiding people, places, activities, objects, and situations that may trigger distressing memories. People may try to avoid remembering or thinking about the traumatic event and resist talking about what happened or how they feel about it.
- **Alterations in arousal and reactivity**: Arousal and reactive symptoms may include being irritable and having angry outbursts; behaving recklessly or in a self-destructive way; being overly watchful of one's surroundings in a suspecting way; being easily startled; or having problems concentrating or sleeping.

I have experienced almost all the various symptoms to a greater or lesser extent, as you can read in chapter 3. However, I do make reservations about the length of time. A month is short, unless, of course, I went on to have PTSD, from which half of adults recover within 3 months, while others continue to experience intrusive thoughts and feelings related to the trauma for many more months and years to come (sometimes even more than 50 years!).

Recovering from a too-long exposure to a too-high level of stress requires a lot of time. The process works like an egg timer. In no time you wind this small kitchen aid up manually, after which it slowly starts to count down. Similarly, stress can spike disproportionally in a fraction of a second during an incident, but returning to the baseline level takes many months. After a day you will have worked your way through the stage of panic, and a few days later you will leave the stage of fear behind you. During the course of the next weeks, the stress level keeps on dropping slowly but steadily. The whole process takes months.

9. DECISION THEORY FOR THE FIRE SERVICE

Incident reports combined with the writings of philosophers of science, neurologists, and psychologists give an interesting insight into the process of decision-making in the run-up to incidents with a fatal outcome. The way our mind filters out sensory cues for further processing and the unconscious mechanisms that determine the cognitive framework or mental model in which all those bits of information will be interpreted are always the starting point. When you intuitively end up in a wrong cognitive framework, the result is a false hypothesis. If that hypothesis deviates too much from what is really going on at the incident ground, a tragedy is looming.

Modern life is played out in dynamic environments that are bursting with stimuli that scream for attention. Our brain simply cannot process the huge volumes of information we are bombarded with all day long. The sensory cues that we take in from the world around us has to be sifted. All the sensations that enter the sensory memory through the sense organs are filtered for the first time based on attention. Only those things to which you devote attention gain access to the working memory or short-term memory, where they are processed into perceptions. You—or, more correctly, your brain—do this automatically and instantaneously. For the mind is the place where past, present, and future meet. Your handling of current situations is affected by experiences from the recent past and predictions or expectations about the near future. Knowledge acquired not only interferes with the perception of the situation you are in, but is also integrated into your judgment about how the situation will develop in the near term. On the basis of experience and expectation, you separate important from unimportant information automatically without effort. The limits of the working memory in terms of how much information it can handle at any given time are the next filter: 60 bits per second, which is equal to the magic number of seven letters or numbers plus or minus two. That is all. How much of this processing capacity is already in use for other conscious thought processes at a given moment limits the amount of additional input you can handle. When workload demands are high, the remaining memory-storage capacity is so small—or even zero when you exceed it—that nothing can enter the mind for further processing. The entrance is simply blocked. Taking in information and processing it requires a lot from our brains. We really do not go through all that trouble to simply ignore that input subsequently.

The interplay of attention and workload on the processing capacity of our brain is known as the *filter effect*. Not only at a fireground but also in our day-to-day life the filter effect can result in a false hypothesis. It can easily mislead us and let us only see and hear what we want to see and hear or make us disregard information to which we ought to have paid attention. This happened on Sunday, March 27, 1977, when in the Canary Islands two Boeing 747 airliners collided on the runway at takeoff. With a total number of 583 lives lost, the infamous Tenerife airport disaster is still the deadliest accident in civil aviation history. The digital flight data recorders and the cockpit voice recorders shed a surprising light upon the causes of that accident.

The day of the collision, it was extremely busy at Los Rodeos Airport (now Tenerife North Airport). At 12:30 p.m. a bomb exploded at a flower shop in the departure lounge of Las Palmas Airport (now Gran Canaria International Airport), and the terrorist claimed to have hidden a second explosive device. Dozens of flights en route to Las Palmas Airport were diverted to Los Rodeos, as were flights KLM 4805 and Pan Am Clipper 1736.

The regional airport of Los Rodeos is not big. It possesses only a single runway to handle both arriving and departing flights. One taxiway runs parallel to the runway. Four exits connect both lanes, but they are not numbered or clearly marked. There is hardly enough space to park the diverted planes on the tarmac. The end of the taxiway is full, and planes also block exits one and two. At 3:00 p.m. the eagerly awaited message is received that Las Palmas Airport is finally to be reopened. Now it is up to air traffic control to get the aircraft to the departure point at the end of the runway. Air traffic control lets the planes taxi in pairs down the runway, where they subsequently are to take exit three or four and continue their way to the departure point on the taxiway. Two aircraft have already taken off when it is the turn of the next pair. Flight KLM 4805 is requested to taxi down the runway and make a 180-degree turn at the end. At the departure point the pilots are to await further instructions. Three minutes later, Pan Am Clipper 1736 may also begin its taxi down the runway, but its pilots are instructed to leave the runway at the third exit and use the parallel taxiway for the reminder of their taxi.

Los Rodeos Airport is situated near the coast at an altitude of 631 meters. At any given moment, cloud banks can drift in from the sea. Perfect visibility can suddenly change into poor. While the Pan Am Clipper 1736 is taxiing down the runway, cloud banks float onto the runway. Because of the poor visibility, the pilots miss the third exit and continue their taxi on the runway. The air traffic controller cannot see the planes through the thick fog. Nor is there visible contact between the jumbo jets. Furthermore, because the airport is not equipped with ground radar, the position of the aircraft can only be established by radio transmission. Let us take a closer look at the exchanged radio messages. For convenience, I have divided that communication into two phases.

Phase One: The Alleged Permission

The world around us influences you and me via the sense organs. We figure out what we are going to do in reaction to those sensory cues and we start to interact with the world. Our brain simply cannot resist interpreting words and gestures of fellow human beings in order to make sense out of them. For this task, it can draw on a rich arsenal of tools. It can jump to a more suitable cognitive framework, fill in blanks at will, interpret ambiguities as a function of our own expectations or desires, and so on. Unfortunately that day the radio messages left plenty of room for interpretation. The filter effect also came into play. "The sooner we get out of here, the better." After an unexpected delay of several hours, that is the mental framework of the KLM captain. His focus is on leaving the airport as soon as possible, and he interprets all the communication in that manner.

As instructed by the air traffic controller, the Dutch Boeing 747 turns 180 degrees at the end of the runway. Then the captain slightly increases the engine power to check the spin-up before departure. The first officer responds and says: "Wait a moment, we don't have ATC clearance yet." The captain replies: "No, I know. Go ahead and ask." As instructed, the first officer contacts the air traffic controller via the onboard radio: "The KLM 4805 is now ready for takeoff and we are waiting our ATC clearance." The aircraft stands still at the departure point on the runway and waits for clearance.

For takeoff, a commercial airliner must get two clearances:

- Air traffic control (ATC) clearance
- Takeoff clearance

Takeoff clearance is the permission to take off from the runway. The *ATC clearance* contains the route the aircraft has to follow within controlled airspace in order to avoid a collision with other aircraft. The air traffic controller replies: "KLM eight seven zero five, you are cleared to the Papa Beacon and maintain flight level niner zero, right turn after takeoff. Proceed with heading zero four zero until intercepting the three two five radial from Las Palmas VOR." Toward the end of this ATC clearance, the captain says "Yes." Apparently he heard the air traffic controller saying something he wanted to hear. As standard procedure prescribes, the first officer reads back the ATC clearance to the control tower. Readback is the technical term for repeating the message by the receiver. It allows the sender to check whether the receiver understood it correctly. But during this readback, the captain increases the throttle, disengages the brakes, and says to his fellow crew members: "We are going." All his actions speak for themselves. The captain thinks he has permission to depart and thus they leave. Meanwhile, the first officer is still reading back the ATC clearance to the control tower. He ends that transmission with the somewhat unusual message "We are now at takeoff" or "We are uh takin' off." What exactly the first officer said is not clear on the tape.

All these communications between the KLM aircraft and the control tower took only 1 minute and 16 seconds, but they are decisive. What could have given the captain the impression that he had permission to take off? To be absolutely clear, he was not authorized to depart. It starts to go wrong when the first officer informs him that they do not have ATC clearance. With the captain's mind set on leaving, that sentence suffices to create the impression that ATC clearance is the only permission still needed. Then the first officer informs the air traffic controller that they are ready for takeoff and await ATC clearance. This radio message is nothing less than an implicit request for two permissions. It consolidates the false hypothesis that one permission suffices for both clearances. The captain's "Yes" on the cockpit voice recorder and the takeoff procedure that he subsequently initiates make it abundantly clear that he has interpreted the ATC clearance in this manner. The use of the word "takeoff" in the ATC clearance enlarges the misconception even further. In your cognitive framework, the filter effect lets you hear (or see) what you want to hear (or see). Here the captain wanted to hear that he was allowed to take off, so he interpreted all ambiguities in that sense.

Phase Two: Missed Ways Out

The unusual sentence "We are now at takeoff" or "We are uh takin' off," spoken by the first officer, is interpreted in different ways by the air traffic controller and the crew of the Pan Am Boeing. The air traffic controller replies with: "Okay. . . . *Stand by for takeoff. . . . I will call you."* The italicized part of this message is barely audible in the KLM cockpit as it is overlayed with a high-pitched squeal. I'll come back to it later. In the mind of the air traffic controller, the Dutch plane is waiting on the runway. The crew of the Pan Am aircraft has another point of view. According to them the KLM Boeing is already initiating takeoff. Immediately after the "Okay" of the air traffic controller, they break into his transmission and say: "No, uh . . . and we are still taxiing down the runway, the Clipper one seven three six." The air traffic controller requests them to inform him when they are off the runway. The Pan Am crew replies: "Okay, we'll report when we're clear." Back in the cockpit of the KLM jumbo jet, the crew could hear the messages exchanged between the American Boeing and the air traffic controller. While they are picking up speed on the runway, the flight engineer hesitantly asks his colleagues: "Is he not clear then?" The captain responds: "What do you say?" The flight engineer repeats his question: "Is he not clear then, that Pan American?" The captain and the first officer respond almost simultaneously: "Sure!" A few seconds later the recording stops with the sound of the crash.

At a certain point in time, the air traffic controller and the Pan Am crew were transmitting messages simultaneously on the same frequency. As a result, crucial information was missed by the Dutch pilots (the italicized part of the air traffic controller's message from earlier). The air traffic controller starts his message with "Okay," which is followed by a pause of 2 seconds. "Okay" is his catch phrase, which only indicates that he has heard the message; English is not his mother tongue. With the subsequent silence, the air traffic controller buys himself a few seconds to formulate an answer: "*Stand by for takeoff. . . . I will call you.*" In the cognitive framework of the Dutch crew, his "Okay" is the confirmation that they are indeed cleared for takeoff. During the short silence of 2 seconds, the Pan Am Clipper 1736 breaks in with the message that they are still taxiing down the runway. The whole time, however, the air traffic controller has been holding down the transmit button, resulting in two transmissions on the same frequency at the same time. The radio interference after the "Okay" causes a high-pitched squeal in the KLM cockpit. Only if you listen carefully can you hear on the tape in a somewhat distorted, though understandable, voice the air traffic controller saying: "*Stand by for takeoff. . . . I will call you.*" But after his "Okay," the attention of the Dutch captain and his first officer are no longer on the radio traffic. It completely shifts to the procedure for takeoff. During takeoff, their workload is so high that in their working memory there is no more processing capacity available for new input. Everything is filtered away.

The control tower's response to the Pan Am crew's message also gets lost, although by then the squeal is gone. Their answer fails to capture the attention of the Dutch crew, as the air traffic controller uses for the first and only time the call sign *Papa Alpha* 1736 instead of *Clipper* 1736. The mental filter of the Dutch crew stood open for Clipper, but not for Papa Alpha. That supposedly other airplane is the least of their worries, so why bother paying attention to that message? When the first officer of the Pan Am Boeing reads back the message of the air traffic controller and says "Okay, we'll report when we're clear," the flight engineer of the KLM plane must have recognized his voice. That is no coincidence. During takeoff, his workload is the lowest of the flight crew. He still has got some processing capacity available. Hesitantly he asks his colleagues whether the American aircraft has left the runway. The quasi-simultaneous "Sure!" of the captain and the first officer leaves no room for doubt. It is only a few seconds later when the Pan Am plane suddenly emerges from the thick fog that the false hypothesis is shattered. By that time and at that speed a collision is inevitable.

After more than 45 years this accident is still food for thought, including for the fire service. During fires, the incident commander usually stands outside while his colleague firefighters are working inside the building. They do not see one another. Information can only be shared over the radio. The view the incident commander has on the incident from the outside can differ substantially from that of his colleagues busy inside the building. For the various parties involved this can lead to different cognitive frameworks. This, in turn, can affect how one and the same radio message is interpreted. The interpretations can even vary to a dangerously wide extent. And then we even assume that the radio message is picked up by the crew inside. In the heat of the moment, many radio messages from the outside might simply not be picked up by the teams working inside, for they are focusing their attention (visual and auditory!) on one task: extinguishing the fire. Moreover, the workload they are facing is at an all-time high. At such moments the crew members inside are unlikely to notice something unexpected, for example, a general call via the radio or the evacuation horn of their fire truck. These warnings fall on literally deaf ears as the crew members suffer from *inattentional deafness*.

What Is Your Impression?

Los Rodeos and the Beringen fire demonstrate that false hypotheses arise easily. But are there ways to puncture those cognitive illusions? We human beings are blind to all pitfalls in our own perception of the situation. We are the least bit aware of the endless stream of assumptions, misconceptions, and first impressions that are irrevocably entwined with our conscious decisions. We hardly question our sense of reality, not only in the run-up to accidents but also throughout daily life. We adhere to one interpretation and neglect to come up with plausible alternatives. Or as D. Kahneman and colleagues write:

> We hold a single interpretation of the world around us at any one time, and we normally invest little effort in generating plausible alternatives to it. One interpretation is enough, and we experience it as true. We do not go through life imagining alternative ways of seeing what we see.[19]

Once you have accepted your own theory (or strategy) and applied it in your reasoning, you become blinded by it, and to come up with new or alternative solutions is hardly possible. Your own theory or mental model is extremely important and remains most influential. Without that framework, you stand nowhere as an incident commander and are bound to be indecisive. You obviously—or better, intuitively—trust your own theory. Precisely herein lies the danger. You do better not to take your own assessment of the situation or strategy at face value. During the incident you have to put it to the test, not once but several times. That is easier said than done. Once you are in the flow of an incident, it is anything but easy to come to an entirely different insight based on a new analysis of the same situation. Do not confuse this new analysis with analyzing new data or thinking 1 or 2 hours ahead in the fire scenario. It is not about that at all. We are dealing with a fundamentally more difficult proposition: *You have to question the sense of reality that you are experiencing at that moment.* Is your interpretation of the events in tune with what is really going on? Are there no other interpretations possible? Colored or clouded by your first impression, your interpretation might not agree to a sufficient degree with what is actually happening. Los Rodeos demonstrates that questioning our sense of reality is extremely difficult. Even the flight engineer's hesitating questions were not enough to raise doubts. When you are so focused on takeoff that you have interpreted everything in that sense, then it is simply inconceivable that the runway might not be clear when you are reaching liftoff speed.

Let us return to the incident in Beringen. That night, the fire truck of Heusden-Zolder arrives at the fireground around 2:50 a.m. My colleagues get out and I lead them to the narrow opening in the back wall of the supermarket (building 6). That is where I want them to enter the building, for they have to locate and extinguish any remaining seats of fire. I tell them what we have done so far: "On arrival we found a blazing fire at the front side. That's extinguished. An additional team was sent into the shop via the right side. They encountered a kitchen fire and have extinguished it. They are now damping down."

That night I did what every fireground commander does at each incident ground: I briefed my colleagues and gave a situation report. We do this to create a shared mental model or a shared representation and understanding of the situation. This all seems more obvious than it really is. As the first incident commander, you lay the foundation for the entire operation. All other decisions will be based on your assessment. You establish the framework wherein the incident will be handled. Little bits of information that you have missed or misinterpreted at the beginning can have a big impact later on. Only after the incident has been dealt with can you know for sure whether your perception of the situation or your sense of reality was sufficiently in accordance with what was actually happening.

But there might be a lot more going on than first meets the eye. What if you miss the ball from the start and slip into a wrong mental model? After all, it is possible that your first impression might be off the mark. Without noticing, you roll into a false hypothesis. Just like at any other incident, you brief your colleagues that arrive later at the incident ground. Before you know it, you end up like *The Parable of the Blind*, as put on canvas by Pieter Bruegel the Elder in 1568. In this painting six blind, disfigured men are walking hand in hand. The leader of the group has fallen on his back into a ditch, and the others will follow suit.

Instead of immediately briefing your fellow officers or warrant officers, I think it is better that you first ask them what their assessment is of the situation. Just like the first incident commander, they also get a first impression upon arrival at the incident ground. Their impression can differ substantially from yours, and this for a number of reasons. The main source of variability within decisions is selective attention. Because of differences in training or experience, your colleagues can have another opinion about the incident. Elements that have fallen into your mental blind spot might well have caught their eye. Or the other way around, they might be able to puncture those sensory cues to which you have assigned too much weight.

The incident itself is another source of variability, surely when your colleagues arrive at the fireground at a later time. Because of the work done by the teams at the incident ground, the initial state that later-arriving colleagues encounter will differ profoundly from that of yours. As the work on the scene progresses, more and more information becomes available. All that extra input greatly augments the possibilities for differing points of view. In the case of a false hypothesis, these are all opportunities to burst the bubble.

Since the Los Rodeos accident, international regulations regarding civil aviation have been stepped up. A new model for cockpit interaction known as crew resource management has been implemented worldwide. In order to reduce human error in the cockpit and improve airline safety, civil aviation goes to great lengths to create an environment in which the captain's authority can be challenged by colleagues in a respectful manner. Crew members are encouraged to voice their concerns freely to the captain, who must hear them out and reconsider a decision in response to the points raised.

Just like aviation, the fire service has an outspoken hierarchical structure and a general culture emphasizing chain of command. Encouraging officers and warrant officers to question the understanding of the situation upon arrival at the scene is the best chance we have to puncture a false hypothesis. The first incident commander must briefly check with them whether the strategy/deployment corresponds to their professional judgment and experience. Premature influence must be avoided. People do not like to disagree, which makes it all too easy to come to a quick consensus about a poor judgment—with all the dreadful consequences this entails, for such a false sense of agreement only reinforces the illusion of an alleged accord. Your colleagues must be able to make their own judgments completely independently. In the next phase, their assessments must be aggregated with that of the first incident commander into a common strategy. The opinion formed independently from you by a warrant officer or another officer is too valuable to be carelessly neglected. The best we can do is to discuss conflicting impressions without getting bogged down in endless talks or—even worse—indecision. The best decisions emerge when teams think together.

As an incident commander you must of course be prepared to adopt this way of working and tolerate what it implies for your authority, for only in this way you can take into consideration your colleagues' assessment of the situation during the decision-making process. You have to give your colleagues the necessary space to allow them to voice their concerns.

Only when your colleagues are feeling secure in your team will they be willing to share their ideas and insights with you. In every team, the triangle of trust, honesty, and self-respect is indispensable for open communication. As an incident commander, you must have confidence in the reports of your colleagues and be willing to believe them and base your decision-making on them. Each team member must report honestly so that all the different observations are available and usable when the decision is made. Everyone furthermore has to respect themselves and their own opinion because each person must share it with a superior or the group, even when knowing that they adhere to a completely different point of view. The basis for the new consensus or strategy has to be a shared belief, not social influence or pressure.

Can the fire service prevent accidents like the one in Beringen from happening again with a better model of decision-making? Unfortunately not: decisions can, and do, go wrong. Accidents happen and some incidents are more risky than others. Something that is exceptionally unlikely or exceedingly rare is more likely to remain unrecognized. And first responders are only human. Bias and noise are omnipresent in human physiology and psychology. They are an intrinsic part of human nature. The mind will always select and interpret information based on acquired experience and expectations. Decisions are also made in the context of the real world, and some contexts or situations are more prone to bias than others. The blazing fire at the front of the derelict supermarket that Sunday night in Beringen was unfortunately such a context.

10. NOTHING HUMAN IS ALIEN TO US

The last psychological element that I will briefly discuss is group dynamics. In the early hours of that Sunday morning in Beringen, no one pulled the emergency brake. That is strange. In total three teams were deployed, each covering another flank. One team attends to the blazing fire at the front. When that is extinguished, another team starts an interior attack via Saint Barbara Street. Later on, a third team is sent in at the back of the building to locate and extinguish any additional seats of fire. Despite the different places of deployment, each team having a different view of the incident, no team suspects that there is more to the incident than a standard ventilated fire.

Before the dramatic turn of events, no one has voiced concerns. No one had any idea what we were in for.

Several factors play a role in group dynamics. Through *pluralistic ignorance* or the *advantage of conformity*, we human beings tend to conform to the rules and norms of our community. We have a natural tendency to do what others are doing. We even give up a right answer or our personal preference when it goes against a clear consensus of the group. All heading in the same direction feels more natural than breaking rank and being the odd man out.

Here the leader comes into the picture as a key figure in groupthink. The more forceful the leadership, the stronger the impact of group conformity. As a rule, a (too) strong leader leaves too little room for others to voice their opinion or propose an alternate solution. A (corporate) culture that is characterized by much respect for authority has the same effect. Between 1980 and 2000, Korean Air experienced a high number of plane crashes because literally no member of the cockpit crew dared to speak up when the captain made a mistake.

In my opinion, the explanation for the group conformity in Beringen lies elsewhere. We may have stumbled upon yet another characteristic of intuitive judgments or automatic mental processes of System 1. Following our (skilled) intuition is not only more natural; it is also more pleasant. The alternative, questioning your intuitive judgments and acting against them, is to be psychotic.

A crucial factor in our group conformity has been the training in the fire service. All firefighters of our Fire and Rescue Service—and by extension all firefighters in Belgium—get the same training and practice. The BE-SAHF model serves as a guideline for reading the fire. However, too little or even no attention is given to the limitations and imperfections of this model. Failing to define a field of application for a purely empirical model is downright asking for problems.

In various incident reports of fires that have claimed the lives of firefighters, a combination of the following critical cues regularly appears:

- Good visibility inside
- Low temperature inside

- Negligible amount of smoke inside
- Smoke light in color
- Smoke that barely flows

In the BE-SAHF model as it stands, these are all cues that are not worrisome for the incident commander who is standing outside, nor for the firefighters working inside the building. The signals suggest a manageable fire of limited power in the growth or decay stage of a ventilated fire development that we firefighters can tackle with our nozzle techniques. And the successes we have repeatedly booked in the past in precisely those circumstances give us the impression that this all makes sense. In the BE-SAHF model, we came to this reading because the model's input included the fire developments that were most common around that time, namely ventilated and underventilated fires.

But today the complexity of the world is growing at a faster rate than ever. The fire service has to deal more and more with construction fires. The cues emitted by this type of fire are precisely the same as those of a ventilated fire, but they must be interpreted completely differently because we are dealing with another type of incident.

In the case of a construction fire, there is no guarantee whatsoever that the conditions are still safe with the aforementioned critical cues. On the contrary, the situation indoors can change completely from one moment to the next, making interior firefighting no longer a safe option. This problem is not a uniquely Belgian one; it is ubiquitous. In the incident report on a construction fire with a fatal outcome in Massachusetts, U.S. (2007), the cues feature prominently: "Upon arrival, fire was showing through the roof with negligible smoke and heat conditions in the main dining area."[20] In the report on the Beringen fire we read: "There was a light stream of smoke coming out of the door as the crew went in. A light grey smoke hung from the ceiling. Its temperature was between 50°C and 60°C."[21] When dealing with a construction fire, the BE-SAHF model deceives firefighters, just as it does with other atypical fire developments like a pulsating fire or a fire gas ignition.

The commission of inquiry that investigated the fatal fire at De Punt in Drenthe, the Netherlands (2008), rightly concluded that every firefighter would have done the same and fallen into the deadly trap. (I am referring to the fire in the boat repair shop discussed in section 6 of chapter 2.) Now what was that deadly trap? A pulsating fire, a particular kind of fire development where periods of a lot of smoke alternate with periods of no smoke.

- Standing in front of the hangar, the firefighters only see a few patches of smoke at the very back of the building before the start of the deployment.
- With this image linked to the information that there is a car burning at the rear left end, they decide to enter the hangar to extinguish the car fire.
- Shortly after the firefighters have entered the hangar, the fire in the indoor storeroom at the back rekindles.
- This time the flames shoot through the wooden roof of the storeroom and ignite the masses of flammable smoke gases that have accumulated in the ridge of the hangar.

In a fire gas ignition in Montréal, Canada (2006), in which one colleague died, firefighters observed the following in the corridor of a residential block: "A stream of white smoke escapes from the upper corner of the door. There is still no heat."[22] January 21, 2006, the Montréal Emergency Centre receives a call for a suspicious odor that is hanging in the third floor of a small flat block containing six apartments. It is 3:28 a.m. According to the caller, it is not a smell of burning, but the odor stings the eyes and makes one nauseous.

At 3:35 a.m., the fire service arrives on-site and members begin to look for the cause of the nuisance. No smoke detectors are sounding, nor does their gas detector go into alarm. But inside the flat block it stinks and the smell takes their breath away. Except for one, all firefighters put on the full-face mask of their breathing apparatus.

The smoke turns out to be escaping from flat number 4 on the first floor. The residents are not in. They are on holiday and have yet to return. The firefighters break the door open. In the flat they try to locate the source of the smoke, and in the meantime they open the windows so the smoke can escape.

At 3:43 a.m. the source of the smoke is found: In the living room, a seat near a window is smoldering. In the incident report, the description of the conditions inside the flat at this time is still the same: "There are no flames, no heat, there is smoke and the smell is still there." The firefighters decide to throw the seat out the window. The cushions have already been thrown out. The seat is next. But when they are manipulating it, a part kindles into a flame and ignites the combustible fire gases that have accumulated against the ceiling. At 3:46 a.m. the entire residential block is ablaze.

The aforementioned construction fire in Massachusetts, U.S. (2007), in which two American colleagues died while conducting an interior attack to extinguish a fire in the roof of a restaurant, deserves a closer look. The fire service wants and has to learn from accidents, but this incident seems to suggest that our eagerness to learn stops at the border. Both the type of building and the fire development are remarkably similar to that of Beringen. It is particularly painful to note that 12 years after the Massachusetts incident, again two firefighters have lost their lives in similar circumstances.

In Massachusetts, a fire was raging in the cockloft of a single-story taxpayer of ordinary construction. Tax avoidance is not only in Belgium a national sport; Americans are also ardent practitioners. In order to evade the tax on vacant building plots, numerous commercial buildings called taxpayers were built by real estate investors from the 1900s onward. Such a commercial building houses several stores under a shared roof. In this case the taxpayer contained four commercial occupancies.

The fire resistance between the various shops is typically minimal. The underlying idea of the taxpayer concept is to house as many stores as possible under a single roof at minimal cost. The roof contains a lot of wood and insulation through which the utility lines run, gas included.

The cockloft is the enclosed space above the finished ceiling and under the roof. As a rule, a cockloft has a thickness of approximately 1 meter and often contains void spaces. The taxpayer building in Massachusetts was an exception to the rule. Here the roof was divided into separate compartments, as the minimal fire resistance of the walls passed through the cockloft. A Chinese restaurant was operated in one of the commercial spaces of this retail warehouse. The total area of the restaurant was 180 m^2 and it was equipped with an open kitchen of 25 m^2.

- On August 29, 2007, a fire starts in the exhaust hood above the cooking stove.
- The fire is reported at 9:04 p.m., and when the fire service arrives at the scene 4 minutes later, the flames begin to shoot through the roof. There is some smoke in the restaurant, and the temperature is negligible.
- In the main dining area, a firefighter is searching for fire extension above the suspended ceiling while an off-duty firefighter arrives on the scene and smashes in the front window of the restaurant.
- The smoke and heat conditions inside the restaurant change rapidly. Pitch black smoke breaks through the ceiling. In no time, the visibility inside the restaurant is reduced to zero and the extremely hot smoke ignites instantly.
- That evening two firefighters die.

The main difference with Beringen 12 years later is that in Massachusetts the fire is already through the roof when the fire service arrives at the incident ground. When flames appear to be shooting through the roof, we are tempted to think that enough of the heat and the smoke can escape. That is not always the case. It is just one of the many assumptions or automatic mental processes of System 1 in the head of firefighters.

Another difference is the size of the building: It is roughly four times smaller than the store expansion in Beringen. The most important similarity, however, is the identical development of a construction fire in a comparable roof construction.

The associative human brain is designed to detect meaning in patterns. It is capable of making sense out of almost anything and it learns from repetition. What do experienced firefighters intuitively think in a building where there is good visibility, light-grey smoke against the ceiling that barely flows, and a temperature of around 50°C to 60°C? In those conditions firefighters not only think they are safe, but they also feel safe.

Evolutionary psychology has taught us that an overly relaxed—in our case, safe—feeling can be dangerous. Emotions that make us feel uneasy often lead to a behavior that is better for us. When perceiving or suspecting danger at an incident ground, fear or another disturbing gut feeling prompts us to become extra vigilant and cautious. Routine and repetition have the opposite effect. Then we are less alert to potential danger and more likely to take risks. All our successes and experiences gained from previous fires play to our disadvantage when we are dealing with an atypical fire development because we are not aware that we are attending to a totally different type of incident. The BE-SAHF model as it stands today only reinforces this automatic thought pattern.

So everything starts with an automatic thought pattern that puts us firefighters on the wrong track, and that is immediately followed by an unconscious behavioral pattern. The following riddle is often used as an example of an automatic or implicit association:

> A father and his son are in a car accident. The father dies instantly, and the son is taken to the nearest hospital. The surgeon comes in and exclaims: "I cannot operate on the boy. He is my son!"[23]

How is this possible? Take a minute to reflect on it. Do you have the answer? A lot of people are unable to answer the question right away. The surgeon is the boy's mother. In our mind, the subconscious association between "surgeon" and "man" is much stronger than that between "surgeon" and "woman" because currently most of the surgeons working in hospitals are men.

An automatic thought pattern or unconscious association within the fire service is that light-grey smoke that barely flows is not dangerous. Another stereotyped association or mental shortcut that played that night in Beringen is linking smoke to a visible fire. First the blazing fire at the front of the building was identified as the source of the smoke, and when that fire was extinguished, the kitchen fire at the rear became the alleged source. In hindsight, most of that smoke can be linked to the construction fire in the roof of the supermarket.

3

A BRAIN ADRIFT

When a fireground commander plots the strategy in a highly uncertain situation under significant time pressure, they do this based on an intuitive or automatic thought process. One of the greatest mistakes of *Homo sapiens* is exalting our conscious thought as the *nec plus ultra*. However, compared to conscious thinking, unconscious thinking is not only much more powerful but is also used more often. We just do not realize it.

In the first half of this chapter, we will take a look inside our head to gain a better understanding of the inner workings of our unconscious mind. In the second half, we will discuss how small disturbances of the delicate electrochemical balance within the neural circuity inside our head cause us to lose grip on reality.

In the months following the Beringen fire, disturbances of precisely those biochemical processes in my brain made it impossible for me to make heads or tails of my thoughts, feelings, and actions. The same mental mechanisms have played tricks on many people and different cultural milieus throughout the history of mankind. The genes inside you and the environment where life is played out dominate your thoughts, feelings, and actions more than you might suspect. The ideal wide-angle lens on all the conscious and unconscious thought processes in our predictive brain is the history of mankind. That is why I have sprinkled this chapter with a myriad of facts and socially relevant themes from past and present.

1. INTUITION AT WORK

On a spring day in 1908, an opportunity arises for the curators of the Egyptian collection of the Museum of Berlin. A Frenchman offers an exceptional artifact from ancient Egypt for sale. The historic importance of the object might well rival the famous Rosetta stone kept in the British Museum, which in 1822 allowed the French scholar Jean-François Champollion to crack the code of the hieroglyphic script.

In book 4, chapter 42 of the *Histories*, the Greek historian Herodotus recounts how the Egyptian pharaoh Nekao II sent out a fleet manned by a Phoenician crew. They were instructed to circumnavigate Egypt, departing from the Red Sea and returning by way of the Mediterranean Sea. The journey took several years and the expedition had to be self-supporting. Whenever autumn came, they had no choice but to interrupt the expedition for several months. They had run out of supplies. Forced to go ashore and roll up their sleeves, they sowed a patch of land and waited for harvest. Several months later, after having harvested the grain, they set out to sea again for the next stage of their trip around Africa. Two years after their departure, they rounded the Pillars of Hercules at Gibraltar, and after 3 years they finally set foot again on Egyptian soil, but this time in the delta of the river Nile.

This wonderful story has exactly the same problem as the exodus of the Israelites out of Egypt under the guidance of Moses. There is simply not one Egyptian source that backs up the account given in the book of Exodus. But that might well change that day in Berlin. On the desk of the renowned German Egyptologist A. Erman lays a copy of a hieroglyphic text of 15 lines. The original is engraved on the underside of a stone scarab or dung beetle measuring 20 centimeters in length. The owner is the son of a well-respected French Egyptologist who died in 1903. His mother told him that his father had excavated the artifact together with another stone scarab in the Nile delta.

If there is one thing that Erman can do better than anyone else, then it is reading hieroglyphs. Thanks to the countless Egyptian texts he is processing for inclusion in the monumental *Wörterbuch der aegyptischen Sprache*, he has developed a sixth sense for the Egyptian language. A quick glance at the

Egyptian text suffices. He is sure: This is a forgery. The text contains grammatical errors that, in his opinion, no Egyptian scribe would ever make. A purchase is not even taken into consideration. That would be the end of the story were it not that a few months later he receives a letter from J. Capart, the deputy curator of the Royal Museums of Art and History in Brussels. His Belgian colleague has purchased both masterpieces for 10,000 francs, a considerable amount of money in those days. Today you can still admire the two oversized stone commemorative scarabs in Brussels.

What do these scarabs have to do with you and me? To be honest, nothing. But the intuitive thought process that made one person buy them and another disregard them is something we use several times a day. In some cases the outcome can even be a matter of life and death, like one morning in Kaprun, Austria.

It is Saturday, November 11, 2000, and Armistice Day is being commemorated. A holiday on a weekend is always nice because, for many of us, it is an extra-long weekend. Today is also the opening day of the new winter sports season, perfect for skiing or snowboarding on the Kitzsteinhorn Glacier in Kaprun. That sunny morning, 161 men, women, and children, together with one train conductor, board a funicular, which will take them in less than 9 minutes' time through the 3.3-kilometer-long tunnel of the strongly sloping Kaprun 2 Gletscherbahn railway.

After 2 years of construction, this masterpiece of civil engineering was opened in 1974. The funicular itself consisted of two linked wagons with a cabin for the attendant at the front and rear. As is typical for a funicular, the wagons were pulled up or let down the tracks by a cable powered by a motorized winch system. There were neither fuel tanks nor engines on board the funicular; they stood in the *Alpinzentrum*, which functioned as the control center and terminal station on the glacier. The supervisor's task was restricted to opening and closing the hydraulic doors, after which the funicular departed. Depending on whether the train was going up the tracks or coming down, the attendant withdrew to the front or rear cabin for the duration of the ride.

In 1994 it was time to give the railcars a major overhaul. Their tatty angular design was exchanged for a futuristic look and feel with elegant curves. This was also an ideal opportunity to address the complaints of the train conductors: They found it too cold in the cabins, as there was no heating in the railcars. That was okay for most passengers because during the short ride the skiers were wearing their ski suits. The conditions were completely different for the supervisors. In wintertime they sat all day long on an icy funicular. An electric heater was installed in each cabin, but instead of a device suitable for vehicles, a fan heater for domestic use was installed under the control panel. The fire extinguishers and emergency hammers on board were locked away. In the attendants' cabins they were of no use to the passengers. The railcars were not fitted with a public address system, nor could the passengers contact the train conductor during the ride. In those days funiculars were considered so safe that all those fancy safety features would just be a waste of money, as having enough lifeboats was on the Titanic.

Just after 9:00 a.m., the railcars packed with skiers and snowboarders plunge into the unlit tunnel toward the *Alpinzentrum*, only to never come out again. The funicular comes to an abrupt stop 530 meters into the tunnel. In the rear cabin, which at that time is unmanned, hydraulic oil is leaking from one of the pipes. Fueled by that flammable oil, the fan heater turns in no time into a serious fire. Temperatures are rising and soon the flames reach the hydraulic lines of the railcars. The brake lines burn through and lose pressure, causing the brakes to snap into safety. The railcars come abruptly to a halt and after the emergency stop, the funicular does not move an inch. And even worse, because of the loss of oil pressure, the sliding doors are blocked.

Now the railcars begin to fill with smoke and a bit later also with fire, fed by the 120 liters of hydraulic oil in the brake lines. After the train conductor seated in the front cabin realizes what is happening inside the railcars, he reports to the control center that a fire has erupted.

The passengers quickly realize what is happening. A dozen of them in the back of the funicular manage to break one of the laminated glass windows with a ski stick and slip out. In the unlit tunnel, they see the flames shooting out the back of the last railcars. When they look up, they see little more than a large black hole. Now they have to make a vitally important choice: Are we going up or down? Going down is not obvious. The service staircase next to the funicular is narrow, and do not forget that there a fire is raging. The heat will be terrible and the fire may even burn your skin. Another option is to calmly go up. Yes, the tunnel with its 41-degree incline is steep, but with every step you take, the distance between you and the fire will grow. Now it is up to you to decide. Which direction are you going? Will you go down past the fire or up away from the fire?

Have you chosen? Okay. Coincidentally, there is a volunteer firefighter with more than 20 years of experience among the 12 people that have managed to escape from the railcars. But more importantly, he can convince the 11 other persons to follow him down. Threatening as the flames and smoke pouring out the last railcar may be, he takes the group in tow and lets them pass by the fire.

Now back to the railcars. These are already so filled with smoke that several passengers lose consciousness. Somehow the train conductor has managed to unlock the doors manually. The passengers that are still conscious can now push the sliding doors open. In no time another group is standing on the service stairs next to the train. Stressed out, they immediately have to decide which direction to take. A crucial difference, however, is that there is no firefighter among them. Their instinct tells them to flee the fire. One person after the other starts the journey up.

All passengers who, like the train conductor, decided to go up died within 142 meters of the train. The only survivors of the Kaprun disaster were the 12 people who went down under the guidance of the firefighter. What did he know that the others did not?

Like every firefighter, he was well aware of the dangers and the behavior of smoke. When there are casualties in a fire, the cause of death is more often the smoke than the flames. Smoke is hot and full of toxic gases, such as carbon monoxide and hydrogen cyanide. Another name for hydrogen cyanide is Zyklon B, the gas used in the gas chambers of the extermination camps of Nazi Germany. One deep breath of hot combustion gases by itself is enough to burn your alveoli and subsequently drown in your own bodily fluids, which start to build up in the lungs (pulmonary edema). And what do hot smoke plumes do? Every firefighter knows that they rise. Translated to the tunnel, the smoke will travel not down but up. In addition, because of its slope, the tunnel will act like a gigantic chimney. Oxygen will be sucked in from below and the choking smoke, scorching heat, and the fire itself will be propelled upward at lightning speed. In total 155 people from eight different countries perished (150 people in the railcars going up, 2 people in those going down, and 3 people in the *Alpinzentrum* on the glacier). The funicular was never reopened.

What does the disaster with the funicular in Kaprun teach us? That day, virtually all people on board the funicular were misled by our innate flight reflex. Like other animals, we humans instinctively flee a fire. The more distance there is between us and a fire, the better. But that is of course reckoned without the smoke. Only the firefighter was not misguided by our innate instinct. Based on knowledge and prior experience he intuitively knew that he had to get under the fire to stay out of the grip of the suffocating smoke. Instead of fleeing up, he went down past the seat of the fire. Compared to the others his knowledge of the behavior of smoke was extraordinary. The more expert you become through years of practice and learning, the better your intuition tends to be and the more unerring it becomes. However, it will never be infallible. Do not follow your intuitive judgments without thinking in situations or circumstances you are not familiar with and when the stakes are high, because then you are just betting on a good outcome.

The firefighter was unconsciously competent in his field. Only when you have reached that stage can you rely blindly on your expert intuition. Unconscious competence is the end point of a learning curve. The American clinical psychologist A. Maslow distinguished the following four stages in his learning model:

1. Unconscious incompetence
2. Conscious incompetence
3. Conscious competence
4. Unconscious competence

Here is an example to illustrate these various stages. Do you hold a driving license like I do? If so, then good, because then at one point or another in your life you have gone through the same stages that I did years ago.

Everything started when I was 18. I needed no car because my bike got me everywhere I wanted to be. If my destination was a bit further away, I rode my bike to the railway station and hopped on the train. I got everywhere I wanted to be, maybe not always on time, but that is another story. I was not aware that I could not drive a car. I did not care. I was unconsciously incompetent.

My sister had got her driver's license when she was 18, so my parents thought I should get mine at the same age. Overnight my situation changed into conscious incompetence. I get behind the wheel of our Opel Kadett and try to depart . . . and the car stalls. I let the clutch pedal go too abruptly and at the same time I did not accelerate enough. Second attempt. Now I am really accelerating and I end up ploughing under the bluestone pebbles of our driveway. That is all you need to realize that you are consciously incompetent and that you will need several hours of driving lessons and a lot of practice before you can drive a car. And let us be fair: At this stage the car is driving you more than you are driving the car.

And then comes the day of the driving test. By now I am convinced that I can drive, although I still have to think every step of the way. But getting that driver's license is not going to be a problem. I thought I was consciously competent. However, things turned out differently. I managed to pass all the tests on the premises of the exam center (e.g., reverse parking, garage in and out, ramp test) with flying colors, but driving the streets was another kettle of fish. According to the examiner I looked far too little into the rear-view mirrors, and on certain stretches of road I had exceeded the speed limit. I was kindly invited for a second attempt. I passed the second test, but again my driving style was commented upon by the examiner: On certain stretches of road I did not drive fast enough!

When you have just obtained your driver's license, you are consciously competent. By getting behind the wheel of a car in a wide range of weather conditions and at various hours of the day and night over the following months and years, you build up expertise and start to develop an intuitive sense for driving a car. To really finish it off, it is best to follow a skid course with the added bonus of a spin on a racetrack, where for once in your life you may take curves at high speed. And now when you get behind the wheel of a car, you are unconsciously competent. You have long forgotten that for departing you have to release the clutch pedal slowly and at the same time accelerate slightly. By now you do it automatically. Every time you fire up the engine, you intuitively feel whether or not you have to accelerate a bit more or less, for example, with a cold engine in a freezing night or an engine still hot in the middle of a summer day. Driving has become an automatic thinking pattern.

Unconscious competence is the stage of reading the fire that every firefighter reaches through the multiple special trainings in the BE-SAHF model and years of hands-on experience at various incident grounds. That is why we firefighters can and may trust our intuitive judgments. We speak about firefighting with knowledge of facts. And then came the fire of that dreadful night in Beringen, the biggest eye-opener in my career as firefighter.

Monday, August 12, 2019

In the morning I can no longer stay in bed. I have to get up and call my warrant officers. I ask them point-blank: “Did we miss something? And if so, what then?” They cannot give an answer. Just like me, they are unable to pinpoint what we have done wrong.

A colleague advised me to take sick leave for at least 1 week. In the morning I go to the family doctor. I tell him about my role in the fateful fire that is now front-page news. He understands the predicament I am in, and I get a week of sick leave. Should I need more, I may always come back.

On the way back home I pass by the burned-out supermarket. I gasp. This is too much for me. Fortunately, my mental state changes immediately. The sad feelings that were threatening to overwhelm me are suddenly swept away by a burst of positive emotions evoked by the flowers laid down in front of the blackened facade. The fire service helps others in their hour of need. Now in our darkest hour, the others are there for us.

Back home, two colleagues pay me a visit. On the terrace we talk about all sorts of things. This afternoon at exactly 1:00 p.m. a minute of silence will be observed in the fire stations all over the country. One of my colleagues asks if I will attend that tribute in Beringen and say a few words. That is not only the last thing I can do but also the last thing I want to do. Now my place is not in the fire station of Beringen. It is in the fire station of Heusden-Zolder. There I will attend the minute of silence.

In Heusden-Zolder the colleagues are surprised by my presence, but they appreciate it. I did not know we were going to wear our firefighting suits for the tribute. “I’ll just watch in plain clothes among the crowd,” I say. They do not want that: “We will find a suit for you.” A large crowd has gathered in front of the fire station. The modest tribute is impressive in its simplicity.

Afterward I stop by the fire station of Beringen. A team has just returned from an incident. They ask how I am doing, and I do what I now can do best: I talk, talk, and talk. Over time, my audience shrinks to three persons. All of a sudden it dawns on me: All three are part of our Fire Stress Team. They are trained to listen and to keep on listening. That is one of the ways they take care of firefighters who are struggling emotionally. At once I break off the conversation so that they can go home. I do not want to be a burden to anybody.

2. THE DISCOVERY OF THE BRAIN

He is completely out of his mind. He does not have two brain cells to rub together. We take it for granted: Thinking is something you do with your brain. In the past that used to be different. *Do what your heart tells you.* If you prefer this way of thinking, then you think like the ancient Egyptians. They thought with their heart.

While visiting a temple library in the northern Egyptian city of Memphis, pharaoh Shabaqa (710 BC) discovers to his great amazement an ancient worm-eaten papyrus. The document relates how the creator god Ptah gave birth to the other deities by means of a thought conceived by his heart and uttered by his tongue. The creation myth impresses the pharaoh, and he orders the text to be copied on a stela. Engraved on a black slab of basalt, the Memphite Theology of Creation withstood the test of time somewhat better than the papyrus because, after years of having served as a millstone, the stela is now on permanent display in the British Museum in London. The hieroglyphic text dating to around 1300 BC tells us how the priests of Ptah thought mental processes worked:

> It came to pass that heart and tongue gained power over all other parts on the basis of the teaching that it [i.e., the heart] is in every body and it [i.e., the tongue] in every mouth of all gods, humans, animals, insects and all living things, the heart thinking and the tongue commanding whatever they desire.
>
> That the eyes see, the ears hear, and the nose breathes air is in order to make report to the heart. This it is that makes all knowledge originate. The tongue it is that repeats what is thought by the heart.

All in all, this is surprisingly modern. Eyes, ears, and nose carry information to the heart, which converts that sensory input into knowledge. And the tongue conveys that knowledge to the world. Even by today's standards, attributing cognition to all living beings is very progressive. A major difference, however, is that all functions that we ascribe to the brain resided, for the Egyptians, in the heart.

The ancient Egyptians had no eye whatsoever for the brain. Nowhere is this more apparent than in the process of mummification. Compared to other organs, the brain did not get a gentle treatment. The Greek historian Herodotus, who visited Egypt around 450 BC, states in book 2, chapter 86 of the *Histories* that during mummification "they first remove the brain through the nostrils, partly by pulling it out with a bent iron, and partly by pouring solvents into the head." Unlike the organs of the abdomen and chest, the brain was not placed in the tomb together with the mummy, probably because it did not emerge from the skull as a whole but only in bits and pieces via the nostrils. Like bodily fluids, the pieces of the brain were gathered in linen cloths and, along with all the material used during the process of mummification, ritually buried in the vicinity of the tomb.

For the ancient Greeks the diaphragm (*phrenos*)—the thin sheet of muscles between the thorax and the abdomen—was the seat of mind and feeling. Their misconception lives on in the modern-day word *schizophrenia*, which literally means "split diaphragm." The ancient Greeks held the brain with its countless folds and creases for a kind of radiator, which served to cool the blood.

By the 13th century AD, human dissections were being performed in Europe, but it was the Flemish Renaissance anatomist Andreas Vesalius who made the difference in his systematic approach. After much nocturnal carving in his cellar, first on looted corpses and later on bodies of executed criminals, he unraveled the human anatomy with an unprecedented precision in the 16th century. The brain did not escape his attention. The seventh and final volume of his *De humani corporis fabrica* (*On the Fabric of the Human Body*), published in 1543, contains several plates depicting an anatomically correct brain. His ideas regarding the working of the brain were less accurate. Vesalius based their functioning on the technology of reference of those days: hydraulics. After all, that science was rather easy to integrate into the then-current theory of the four humors or fluids of the second-century Greek physician Galen, whose teachings were held as gospel in the study of medicine until well into the 16th century. Thus Vesalius searched in vain for the flow routes that funnel "brain phlegm" into the nostrils.

What René Descartes, the most influential philosopher of the Age of Enlightenment, in the 17th century wrote about the relationship between matter and mind is a completely different story. Even nowadays, Western sciences and humanities find it hard to get out from under the long shadow of one of his few mistakes. Descartes is famous for the aphorism "I think, therefore I am." If there is one person who has put our conscious mind on a pedestal, then it is surely this French philosopher, but unfortunately our unconscious is still paying a heavy price for it. Descartes was of the opinion that matter and mind are two totally distinct substances. The division between the fleshy body (including the brain) and the immaterial spirit capable of reason is sharp, and the difference in their nature unsurmountable. In his *Discours de la méthode* (*Discourse on the Method*), published in 1637, he puts it as follows:

> From this I knew that I was a substance of which the whole essence of nature is simply to think, and which, in order to exist, has no need of any place nor depends on any material thing. Thus this I, that is to say, the Soul through which I am what I am, is entirely distinct from the body, and is even easier to know than the body, and even if there were no body at all, it would not cease to be all that it is.[1]

Thus Descartes claims that body and mind are totally distinct entities and that our immortal soul does not need a body to exist. Our body as a biological organism is subjected to the laws of nature, but the immaterial soul or mind with its refined functions of thought and morality is a completely different substance. Every search for the seat of consciousness inside the body is futile.

He furthermore expressed the view that "there is nothing in the mind of which we aren't aware." In the *Objections and Replies* to his *Meditationes de prima philosophia* (*Meditations on First Philosophy*) from 1641, he dwells on that statement and writes:

> It seems to me self-evident that the mind, considered as a thinking thing, can't contain anything of which it isn't aware. We can't make

> sense of the proposition that the mind, seen as a thinking thing, contains something that isn't a thought or something dependent on a thought . . . and we can't have any thought that we aren't aware of at the very moment when it is in us.[2]

In this way Descartes made the conscious mind more important than it actually is. For this reason, still too many of us underestimate the importance of the unconscious.

Around the end of the 19th century, the idea that the unconscious mind must be not only much larger but also more powerful than its conscious counterpart was gradually gaining foothold. But that progress soon derailed completely and the unconscious mind was demonized as a medieval dungeon packed with suppressed, primitive, infantile, animalistic urges and repressed traumatic memories. This reputation damage is due to the father of psychoanalysis, the Austrian neurologist Sigmund Freud.

However, Freud did publish several general theoretical ideas about the unconscious mind that are rock solid. "Every mental process begins unconsciously, and it can remain unconscious or progress into consciousness," he says, "depending on whether it encounters opposition or not."[3] Unfortunately, those sound theoretical ideas have largely been overshadowed by the work for which Freud has become known, or shall we say infamous: his clinical research on the unconscious or psychoanalysis. For example, in the treatment of obsessions and compulsions, the breakthrough came only after his time-consuming and unsuccessful method of psychodynamic psychotherapy ("the insightful treatment") was abandoned. Psychiatrists and psychologists have stopped searching for the deeper meaning and the underlying, unfulfilled (sexual) desires of an obsessive-compulsive neurosis through countless, endless talking sessions. With psychotropic drugs, behavior therapy, cognitive therapy, and, when necessary, neurosurgery better results are achieved.

We owe to Freud's psychoanalytic work the stereotyped dichotomy of the brain that is all too often still proclaimed: The smart, important consciousness is the master of our brain, and it restrains all our primitive urges that arise out of our stupid, unimportant unconsciousness. And that is all for the better, because otherwise life on earth would be one great Sodom and Gomorrah!

Tuesday, August 13, 2019

Day two of my sick leave. Sitting at home is nothing for me. I am and I feel far too restless. I have to get out of the house. The sooner the better. In the morning I go again to the fire station of Heusden-Zolder, this time to put forward my theory about the distinction between a *fire in a building* and *a building on fire* to an instructor of the fire academy. In the fire station I talk to just about everyone except for the instructor I have arranged to meet.

I encounter the colleague who addressed me that fateful night about the water supply for their fire truck. He makes my day by sharing precious bits and pieces of his experience. He describes the atmosphere in the fire truck of Heusden-Zolder when they arrived at the scene: "Those guys from Beringen have fooled us. For once we may offer them support, and now they have already extinguished all the fire themselves. We are going to be sent back by the officer, or asked to help coil the hoses." Even more valuable is his addition to the order I gave at the backside of the supermarket when I asked his colleagues to enter the building and find and extinguish any additional seats of fire. I concluded my instructions by saying: "But don't go too far into the building."

The afternoon is windy and bleak. I drop my sons Jonathan and Matthias at the swimming pool of the leisure site be-MINE in Beringen. Without a doubt, they will enjoy themselves all afternoon long. Meanwhile, I will try to clear my head and take a long stroll on the big slag heap nearby. During the climb to the top my smartphone rings. The wind blows so hard into the microphone that the human resources director barely can hear me. She lets me know that the fire service is offering me professional psychological counseling. I have already discussed that with my wife. During the visit that the head of the operations department paid us on Sunday evening, she mentioned it. Her position was clear: "You do it. End of discussion." I was less convinced. "I am not crazy!" In order to keep the peace, I promised to accept all help that is offered. Even if it does not work, it does not harm. Suddenly this all becomes very concrete. The psychologist drops by our house that evening.

I am impressed by the psychologist. By this time I have become a motormouth. Once I start talking, I simply cannot stop. "You are a talker," she says, "but talkers have difficulties with emotions." Thus she characterizes me at the end of the intake interview. In the first part she has kicked in an open door, but in the second part she hit the nail on the head. I say to myself: "Luckily I didn't ignore the advice of my wife. This might help."

3. THE CONSCIOUS AND UNCONSCIOUS MIND

The Dutch psychologist A. Dijksterhuis casts light on the perplexing inner workings of our brain in his book *The Smart Unconscious* (*Het slimme onbewuste*). I draw heavily on it. First and foremost, we must get rid of all our stereotypes about the brain. As high as our consciousness is praised as the crown of human reason that separates us from other animals, so unjustly is the unconscious mind belittled as primitive, stupid, and unimportant. In this way we give our conscious awareness more credit than it is entitled to.

The evidence is compelling. Experiment after experiment shows that our conscious mind always lags behind. Using the technique of electroencephalography (EEG), neurophysiologist B. Libet demonstrated that our unconscious is already preparing each action or behavior about 800 milliseconds in advance (almost a full second!) while the conscious decision is recorded only 350 milliseconds before the action. All conscious behavior is prepared unconsciously, and our subjective experience of the conscious decision to act is little more than a sham decision. However, our conscious mind does a great job in hiding the fact that it misses the start over and over again by creating the illusion that it is at the origin of all our decisions and subsequent behavior.

Often we do not know why we make certain choices or do certain things. For example, if you put four identical pairs of nylon stockings in a row, 40% of research participants will take the far-right pair. Should you ask them why they have chosen that particular pair of stockings, they will answer that those are more beautiful, the texture is better, or the color nicer. Based on countless tests, psychologists are convinced that we make a decision or have an opinion on the basis of unconscious reasons, after which our conscious mind starts to rationalize that judgment and comes up with good reasons for it.[4] Or as J. Bargh writes: "In many cases it's the conscious mind that plays explainer afterward, trying to make sense of a judgment that we seemed to already 'know' so solidly that our assessment felt like an incontrovertible fact."[5]

The most striking difference between the conscious and unconscious mind is their processing capacity, whereby our consciousness stands out in a negative manner. Seven letters plus or minus two, and that is all. The processing capacity of our conscious mind is 60 bits per second. Attention is indeed a limited resource. That is the reason you can only pay attention to one thing at a time. You simply cannot read these lines and at the same time draw up your weekly shopping list. The conscious mind works serially: You have to stop one task before you can start another. Switching between various tasks is something you do quickly and constantly all day long.

The unconscious, on the other hand, can perform multiple tasks at a time, and it multitasks throughout the day. While you are reading these lines, not only do your temperature and blood pressure remain constant, but you also do not stop breathing and you are unconsciously aware of your surroundings. In the background you can hear the clock ticking or the dog barking in the garden. Your unconscious can run all these automatic operations in parallel because its computational power is no less than 200,000 times bigger than that of your conscious awareness: about 11.2 million bits per second.

What is inside our unconscious? Reflexes, intuitions, and automatic thinking patterns. Reflexes are innate behavioral patterns that are transmitted through DNA. Intuitions and automatic cognitive processes, on the other hand, are unconscious modes of thinking that are nurtured and reinforced by training, knowledge, and experience. The more you focus on a particular field of study, the better your unconscious mind becomes at it. But the unconscious mind also absorbs the various stereotypes and biases that thrive within our culture, regardless of whether you adhere to them or not. Through selective information reduction and the *dispositional representations* in our associative brain, stereotypes and biases light up in the mind at lightning speed. They act like cognitive shortcuts or neurochemical algorithms that make you ascribe all sorts of alleged qualities to things or persons based solely on the group to which they belong. First impressions and intuition work on this principle.

The fascinating book *Blink* by the Canadian journalist and writer M. Gladwell is devoted to the miracle of intuition. In the epilogue he recounts how his repeated contacts with law enforcement during his midlife crisis led him to write the book.

Men and hair have a special bond. One guy runs into trouble because he has too much hair, while another one's problem is that he no longer has any. Gladwell belongs to the first category, and slowly but surely I am becoming eligible for full membership of the latter group. In his forties, Gladwell began to long for the savage locks of his teenage years. As his hair grew longer and more wild, his life changed in a small but significant way: His contacts with the strong arm of the law increased markedly. He was regularly pulled out of airport security lines for closer inspection. On an overaged hippie each customs officer expects to find at least some amount of cannabis. An all-time low was when he was mistaken for a rapist by three police officers in mid-town Manhattan. All he had in common with the sketch and the description of the 15-year-younger suspect they pulled out was a large head of curly hair.

What Gladwell experienced were, all in all, trivial misunderstandings compared to the past when such social stereotyping managed to take deep and widespread roots in the collective unconscious, often resulting in countless innocent victims. The boundary with stigmatization is subtle.

For example, were you in the 16th or 17th century in the Low Countries an old, poor woman who lived in isolation on the outskirts of a small town or village together with her cats? And did you have a more than average knowledge of (medicinal) herbs and plants? Then you ran the risk of being sentenced to be burned at the stake as a witch. A radical change in attitude toward the rural magic of male and female healers is at the heart of countless witch trials, which from 1530 onward were only held in civil courts and made many victims. The unconscious opinion about these single women and occasionally men had turned extremely negative.

In 2024, society is still struggling with this stereotype and prejudice. Little has changed. *Marginal* still evokes the implicit and unconscious prejudicial association *worthless*. When you are living on the margins of society, you are at greater risk to be treated roughly by your fellow man. Just read *The Book of Daniel* (*Het boek Daniël*) by the Flemish writer Chris de Stoop. In 2014 his somewhat eccentric uncle, who lived as a recluse in his courtyard farm among the fields, was slaughtered in an incredibly brutal manner by some youngsters for a mere 1,000 euros. Whether it be 16th-century witch trials, discrimination against minority groups, or blatant racism today, negative unconscious opinions direct our behavior more than we are aware of.

The brain can best be seen as a group of separate modules (e.g., perception, emotions, language, coordination of the body, higher cognition) that to a certain extent operate independently from one another. In a somewhat disorderly but quite efficient way, they know how to work together.

The cognitive neuroscientist B. J. Baars compares the immensely complicated inner workings of the human brain to a theater. Each theater has a stage with actors, a spotlight, stage technicians working behind the scenes, and a vast, silent audience seated in the theater hall. When the lights dim in the hall, the audience turns their attention to the actors on the stage as the stage technicians put the spotlight on them.

- Our working memory or consciousness is the *stage* of our brain. It is remarkably limited and operates serially: one thing at a time. As this workspace is brightly lit, consciousness offers our sensations, perceptions, thoughts, and sentiments the unique opportunity to widely disseminate their information to the members of the *audience* in the darkened theater hall.
- The *audience* represents the exceedingly complicated behavioral patterns, automatic mental processes, language structures, and memories in our unconsciousness that, like visitors in a theater, cannot see each other in the dark.
- The *actors* on the stage are the wide variety of sensations, perceptions, thoughts, and feelings we experience day in, day out. Numerous sensory stimuli and ideas can reach consciousness at any time of the day, but we can devote attention to only one at a time.
- Just like a *spotlight*, our attention determines the content and direction of our consciousness. Our intentions, expectations, assumptions, and implicit associations shape our subjective experience. Behind the scenes they perform tasks similar to those of *stage technicians.*

Let us now suppose one of the modules of your brain suddenly spots that something is going wrong and urgently needs the other modules to solve the problem. In Baars's theater metaphor, the stage or our working memory is the central workspace in our brain where that module can inform the vast audience of modules in our unconscious mind about the problem. An example for clarification: Imagine you are cycling to the bakery for a loaf of bread, but by mistake you have turned down the street of the butcher.

- The module that monitors whether you are achieving the set goal suddenly notices the error and alarms your conscious mind that you have taken the wrong road.
- Thanks to consciousness, all other modules are immediately informed and take the necessary measures to rectify the error.
- More concretely, you stop, turn your bike around, cycle back up the wrong street, and take the right road to the bakery.

Where exactly is our consciousness located inside the brain? That quest we have to abandon, but not because the mind is immaterial and immortal, as Descartes claimed. Out of the vast network of many billions of neurons or nerve cells and many trillions of *synapses* or connections between them, (self-)awareness came into being in one way or another. Exactly how this occurred is not (yet) known. So many different brain regions are in one way or another involved with our conscious awareness and the creation of thoughts that this is more likely the result of the manner in which brain activity takes place rather than the location wherein it takes place. The Australian philosopher of science P. Godfrey-Smith puts it as follows: "Brain processes are not causes of thoughts and experiences, they are thoughts and experiences."[6]

With regard to the usefulness of our conscious mind, the Dutch psychologist A. Dijksterhuis is quite clear: "Most researchers, such as psychologists and neuroscientists, assume that conscious awareness probably has some use, although it is still quite unclear what use that should be."[7] Many scientists do not hesitate to go a step further and boldly state that consciousness is, from a biological point of view, a useless byproduct of certain cognitive processes that even causes mental pollution. This does not, however, distract from the fact that our conscious mind can have great moral and political value.

Does this imply that the curtain has fallen on our free will and that there is nothing more in our brains than a vast collection of biochemical algorithms, as bioscientists claim, or an unconscious will, as neurologists argue? The final word about free will has not yet been written.

Still, all the subjective experiences that you and I experience every day that let us enjoy anything and everything are by far the most pleasant side effect of our conscious awareness. A good book, a nice lunch with your loved one, a nice day out with your family, or—better yet—a beautiful holiday: it does not matter. Useless as all these subjective experiences of yours and mine may be in the greater scheme of things, these are the very things that bring color to life. For that reason alone you would not want a life without conscious experiences!

Wednesday, August 14, 2019

After 2 days of sick leave I can no longer stand it at home. Not that I have spent much time at home. It is time to get back to work, and that is a good thing. In those days, not much work is being done in the fire station of Beringen, but a vast amount of sentiments and emotions are being processed. Almost every firefighter on duty that night drops in for a chat. The long talks around the kitchen table help tremendously. I am pleased with their input, and hopefully they will gain as much benefit from my input.

Although I dread it, the time has come to return to the fireground. The overwhelming impression that the horrific building made on me when I inadvertently passed it 2 days ago on my way back from the family doctor is undoubtedly at the heart of this aversion. My colleagues do not have the time to accompany me, so I go by myself. In the parking lot of the supermarket I meet a warrant officer from Hasselt, who has come to the fireground to gain a better understanding of how the events unfolded that night. I am pleased I do not have to face my nightmare alone.

At the initiative of the Fire Stress Team, all members of the fire service, the emergency medical service, and the city of Beringen deployed the night of the fire will gather this evening in the fire station of Heusden-Zolder. The Fire Stress Team offers psychological assistance to first responders involved in exceptional, traumatic, or shocking incidents. After all, firefighters and paramedics are not made of stone. The work the Fire Stress Team has done voluntarily and unconditionally in the aftermath of the Beringen tragedy is invaluable.

The meeting runs an hour over time. At the end everyone is given the chance to air their feelings. No one lets the opportunity slip. Nice things are said, and I get the present of my life. It is the turn of one of my warrant officers. Without the slightest hesitation he says that he stands a full 100% behind all orders given by the officer. Spontaneous applause erupts.

A bit later it is my turn to say a few words. For days I've done little else but talk, but now I am somewhat lost for words. I am thrown off balance by what my warrant officer has said. I put up a story about the temporary tattoos that the Fire Service Association Flanders (*Brandweer Vereniging Vlaanderen*) has handed out in the run-up to May 4, the International Day of the Fire Service. Did not on one of those tattoos stand the text "Proud of my firefighter"? I make it abundantly clear that I am proud of every firefighter in the room, and for me each person in the room is a firefighter.

4. HANDLING AN INCIDENT WITH THE HANDBRAKE ON

Now it is time for an experiment. You are about to take a short scrambled-sentence test. Below you will find ten sets of five words. Your task is to make an intelligible four-word sentence from each line, and you have to the time how long it takes to finish the task. Are you ready? Start the clock and begin.

1. him was worried she always
2. from are Spain oranges temperature
3. ball the throw toss silently
4. shoes give replace old the
5. he observes occasionally people watches
6. he will sweat lonely they
7. sky the seamless grey is
8. should now withdraw forgetful we
9. us bingo sing play let
10. sunlight makes temperature wrinkle raisins

Everything solved? Stop the clock. You have earned yourself a reward. You may take a soft drink out of the fridge or make yourself a nice cup of tea or a coffee. Let us return to the test. What do you think of it? It looks like a run-of-the-mill language test. How long did it take? To be honest, your time does not matter. I just wanted to throw you off track and create time pressure. The test is all about your unconscious. Each list contains an innocuous suggestion of old age: *worried, silently, old, lonely, grey, forgetful, bingo, wrinkle,* and *Spain*, the country where many Belgian seniors prefer to spend the winter season. While you thought you were taking a simple language test, the test is designed to let your adaptive unconscious reflect on what it is like to be old.

"Now, wait just a minute," you might be thinking. "Are you saying that you have affected my unconscious with those 10 sentences?" Well, with this test the American psychologist J. Bargh demonstrated the concept of priming.[8] *Priming* is some kind of preprogramming of the unconscious mind. His book *Before You Know It: The Unconscious Reasons We Do What We Do* is not only brilliant but also a fascinating introduction to this and numerous other hidden mental mechanisms in our unconscious mind that secretly govern every aspect of our behavior as we go about our daily lives.

The test you just took is a Flemish adaptation of one of Bargh's renowned experiments. For that experiment he primed half of his test subjects with these stereotypes of old age. After the test, they were asked to go from the office on the first floor where they took the test to the seventh floor for the next part of the test. The test leaders, however, secretly timed how long it took each research participant to walk from the test room to the elevator. One of the stereotyped traits about seniors is that they are a bit slower. Now what were the results? It took all participants primed by old age a bit longer to get to the elevator. Regardless of age, their unconsciousness made them walk slightly slower to the elevator than their nonprimed peers. And you? What did you do? I suggested that you get yourself something to drink after the test. Did you rush or shuffle to the refrigerator? My bet is on scenario two.

It is crucial to realize that stereotypes are central in priming, as is the case in the overall working of the unconscious mind, regardless of whether the stereotype is "true" and whether you believe it or not. Everything revolves around the traits that we ascribe to certain groups or items in our culture. The role the mass media plays in shaping all those various cultural stereotypes can hardly be underestimated. All in all, several types of human behavior are easy to affect by priming with all kinds of traits and stereotypes. It can even make you change your mind. Subsequent studies have shown that test subjects primed by old age, for example, are more likely to agree with conservative statements (e.g., that there is an overload of sex on TV). Even the drive to perform and the goals of others can affect us subconsciously in this way.

Thinking that we can thus completely reprogram the unconscious mind through intensive brainwashing is an illusion. Firstly, priming is one of the most subtle strategies of social influence. Once you are aware that someone is trying to pull your strings, it no longer works. The effects are also short-lived. The pace of the test subjects to the elevator might have been just a tiny bit slower, but once outside in the parking lot they walk again as fast as before to their cars. It is not possible to change permanently the norms and values of a person by priming. We do not notice it, but our judgment and behavior is influenced throughout the day by all kinds of stereotypes and environmental factors. It is precisely because of this uninterrupted, subtle influencing by the things we think of and the people around us that we are little more than human ping-pong balls. On the other hand, it is precisely this internal trait that makes us human beings the hypersocial animals we are.

Firefighters are not immune to priming during incidents. I have always found it peculiar that firefighters behave completely differently during a hazardous material incident than during a fire. It is as if we attend to a hazardous material incident with the handbrake on. One of the challenges you face as an officer at a fireground is keeping your men out of the building: "At last there is a fire to attend to and we are firefighters, so now we want to and will extinguish the fire."

This eagerness is almost completely absent in the event of a hazardous material incident (e.g., a leak in a chemical factory), and even more so when radioactive material is involved in the incident. Then firefighters are reluctant, and in one way or another the fireground commander will have to prompt them into action. Could this behavior have its origin in priming of our adaptive unconscious by the word "hazardous" in the emergency call for a hazardous material incident?

In the difference of approach between a fire and a hazardous material incident, you clearly see the impact that the perception of a problem has on the behavior of the firefighters on scene. The difference in safety margin that the fire service applies to interior firefighting compared to hazardous material incidents is striking. In the event of a hazardous material incident, the safety margin within which the fire service operates is sharply defined. At 10% lower explosive limit (LEL) we are 90% away from the LEL of a flammable gas mixture, but the alarm signal of our explosimeter goes off for the first time. The combustible gas detector reminds us to reflect carefully on what we are doing. At 20% LEL we are still 80% away from the LEL and the alarm of our explosimeter sounds for the second time. The team inside must immediately abort their mission and retrace their steps. The fire officer will have to come up with a safer way to tackle the problem.

And what does the same fire crew do when they attend to a fire inside a house or a building? In what way does the smoke they encounter during their offensive interior attack differ from the gas that is released during a hazardous material incident? Basically none. Fire gases are not only always toxic but also flammable. The only difference is that for interior firefighting we do not have a gas detector at our disposal similar to an explosimeter that warns us when a predetermined alarm threshold is exceeded. The firefighters assess the situation to the best of their abilities with a thermal imaging camera and the effect of the gas cooling, which they perform with their nozzle as they advance inside the building. We use 300°C as the upper limit on our thermal imaging camera. We really do not want to be and should not be in those compartments. But when, for example, we measure 280°C in a certain room, are we above or below 20% LEL? We simply do not know, but interior firefighting is our core business. If there is one thing that we firefighters can assess well, then it for sure must be this. Or at least that is what our consciousness tricks us into believing. In interior firefighting, we have indisputably worked more often above 20% LEL than below. But because we cannot measure the danger with an explosimeter and we tend to grossly overestimate what we think we know about fires, our risk assessment is considerably lower.

And fire officers, are they perhaps immune to priming? On the contrary. We plot the strategy under the unconscious influence of—you already guessed it—priming. Section 5 of the previous chapter has the title "To Recognize Is to Decide." It is the eloquent translation of the principle that, according to the American psychologist G. Klein, is at the heart of our decision-making: recognition-primed decision. A less fluent but more accurate rendering is "decision based on recognition under the influence of priming." Precisely because our strategy is dominated by our first impression upon arrival, our decision is made under influence of priming by all the stereotypes and biases about fire and firefighting that we as incident commanders have built up over time in our unconscious mind through training, knowledge, and experience.

And thus something that began as a harmless psychological experiment that made you walk just a little bit slower to the refrigerator has exposed one of the great unconscious forces that not only has more impact on our firefighting strategy than any incident commander wants to believe but also shapes our behavior all day every day.

Sunday, August 18, 2019

The funeral service was held yesterday. It was emotionally heavy. To honor the deceased firefighters, there is a silent walk in Beringen this afternoon.

Like every Sunday morning I go for a run. My youngest son Matthias accompanies me on his bike. I do not think he does it because he wants to but because he has to. I notice that the boys behave differently toward me. I have the impression that Mama has asked them to keep a close eye on me. This is the world turned upside down. Parents watch over their kids and not the other way around. I am anything but pleased with the situation I find myself in. I want my old life back, the sooner the better!

When we leave the house, it is raining a bit. By the time we have reached the forest, the floodgates open and it is pouring down. It feels great. The rain washes the burden of the past days away. The puddles on the forest trail grow visibly larger and larger. The forest trail becomes one large puddle of water. All my misery lies on the ground. I enjoy running through it and I just cannot stop. I feel liberated, at least for a while.

Matthias starts to get cold and wants to return home. Reluctantly I stop one lap earlier than usual. On our way back we have already passed the football field when I ask him to just follow the road and wait for me when he has reached the canal, while I take a forest trail. He has barely left when fear grips me. *What have you done now? Suppose something bad happens to him! How could you ever forgive yourself?* I must set it straight right away. With some shouting and running faster we are in no time back together. We agree to stay together and not lose sight of one another.

The silent walk starts at 2:00 p.m. at the fire station of Beringen. The turnout is tremendous. My sons, wife, and parents join the walk. The crowd slowly starts to move, and in a serene procession we walk to the incident ground.

A crowd awaits us in front of the blackened facade of the burned-out supermarket. Altogether we are an impressive group about 800 strong. A few texts are read out. They are personal and sober. The ceremony is over. The lady that has organized the silent walk stands next to the microphone. I must thank her. I go to her and give her a big hug. The hug I get in return is even bigger. The hug of my life.

5. A THEORY MORE OR ONE LESS?

It was for sure not the best day of his life when the German Egyptologist Erman found out that the Royal Museums of Art and History in Brussels had bought the commemorative scarabs of pharaoh Nekao II. What did he overlook that had caught the eye of his Belgian colleague Capart? How had he let slip through his fingers the opportunity to buy two masterpieces for the Egyptian collection of the Museum of Berlin? Was he misled by his intuitive judgment, and should he have taken a closer look at the hieroglyphic inscriptions on the scarabs? Yes, they contained prepositional phrases that were strange, but what if this was the first time that such a grammatical construction was found in an Egyptian text? What if the scarabs were black swans, which he had found not white enough? Before Captain Cook set foot on Australian soil in 1770, all swans were white by definition. In the ornithological description of this elegant water bird, the color white was one of its defining features. The European ornithologists did not know better and could not know better, for all swans in the then-known world were white. You can imagine their surprise when suddenly on the other side of the world black swans turned up. There must be a first time for everything, including ancient Egyptian prepositional phrases. Erman decides to take a closer look at the hieroglyphic inscription on the scarab, but this time he pulls out all the stops.

And what would not have crossed the mind of his Belgian colleague Capart before buying the two masterpieces? Capart's knowledge of the Egyptian language was no match for a scholar of great stature like Erman. But other esteemed Egyptologists whose knowledge of ancient Egyptian could rival that of Erman did not have the slightest doubt about the authenticity of the hieroglyphic inscriptions on the scarabs, so why should he have second thoughts? And then there are the scarabs themselves. It is rather easy to link them to the issue of commemorative scarabs that flourished under pharaoh Amenhotep III (1370 BC).

For Amenhotep III, any reason was good to issue a commemorative scarab. Political propaganda is something you have to take care of. A memorial scarab is similar to a contemporary commemorative medallion. These scarabs in blue or green steatite measure between 5 and 11 centimeters. In total, some 200 of these scarabs have been preserved. On the basis of the hieroglyphic inscription incised on the underside, you can divide them into three groups:

- Hunt scarabs are the first group. No bear was safe from the ardent but lousy hunter that the Romanian dictator Nicolae Ceaușescu was. Likewise, Amenhotep III shot many a wild bull (106 in total) and many a lion (102 in 10 years' time) "with his own arrow" during royal hunting parties. Or at least that is what he boasts about.
- Another group deals with feminine beauty. One category relates the arrival of the Mitanni princess Gilu-Khepa with her retinue of 317 ladies-in-waiting into the pharaoh's harem. In those days the Mitanni, just like the other empires of the Near East, sent princesses from northern Syria and Iraq to the Egyptian court as diplomatic brides for the pharaoh. In the other category of this group, the pharaoh presents us the parents of Queen Tiye.
- In the third and final group, Amenhotep III boasts of the immense artificial lake of 3,700 cubits × 600 cubits (roughly 1,924 m × 312 m) that he had dug out for his chief wife Queen Tiye near his palace complex on the western bank of the city of Thebes. It is amazing what length a pharaoh went to just to impress his favorite wife!

More than 750 years separate Nekao II from Amenhotep III, and not a single commemorative scarab is preserved from that intermediate period. Is that a problem? Not at all; anyone looking for confirmation for a theory will always find something. That is what confirmation bias is all about. Nekao II was a pharaoh of the 26th dynasty or the Saite Era. For what is that period known? The Saite Renaissance! Let this be a period in Egyptian history where the pharaohs in art and culture strongly hark back to Egypt's illustrious past. The two memorial scarabs of Nekao II fit perfectly in what

Egyptologists know—or think they know—about Egyptian history. But do they fit into the history of Egypt itself?

I take the actions of Egyptologists Erman and Capart as examples of something you and I also do regularly. When an intuitive belief or an unconscious idea pops into your conscious mind or working memory, you become aware of it and start to reflect on it. As Dijksterhuis says, "When you think, you start to use (in part) different arguments than the ones you would use without thinking, and those other arguments can make for a worse choice."[9]

The American psychologist J. W. Schooler gave this effect the name verbal overshadowing. In his opinion, the left hemisphere of our brain thinks in words and the right in images. I think that is unlikely. I prefer the hypothesis that our working or short-term memory is mainly auditory and focuses on processing information in the form of words and specific sounds. After all, we think in sentences and words (inner speech) and not in images. When you are thinking, you can hear yourself talking in your head but you do not see a movie on a screen. Or at least I do not. The long-term memory, on the other hand, relies heavily on visual and semantic qualities. When you have to account for an intuitive judgment, you can only do that in your working memory or conscious mind with words. At that very moment you let go of your intuition. I'll make it less abstract with a few examples.

Imagine, if you will, the following scenario. Recently you have dined in a nice restaurant where you were served by an excellent waiter. You were so pleased with the service that you gave him a big tip. Now we turn to a scene that is inevitable in every police movie: A number of suspects are lined up behind mirrored glass and you are asked to point out that waiter. You recognize him without a problem. Now imagine the same scenario again, but before the lineup a police officer wants you to write down as accurately as possible the facial features of the waiter. What was the color of his hair? Did he have a small scar on his face? And what clothes did he wear? The more detailed your description, the better. When you subsequently have to pick that person out of the row of lined-up suspects, you are more likely to misidentify him. Writing down a description of his face has impaired your otherwise effortless ability to recognize him later. While you were noting down what the waiter looked like, your initial visual perception was overshadowed.

This made the second lineup more difficult because that identification was based on the memory of what you wrote down and no longer on what the waiter looked like. And we all recognize a face far better from visual imagery stored in long-term memory than from a verbal description produced by our working memory.

The problem with thinking a lot, or even too much, is that arguments that can be verbalized or easier put into words become more consequential than they actually are. Emotions and feelings play a crucial part in important decisions, but you cannot verbalize them well. You can undoubtedly tell me in great detail why you have recently bought a specific car model, in that color, of that make. But can you tell me with the same ease and fluency why you love your partner? It is not so hard to come up with some nice qualities of this person that actually do not matter and for sure do not convey your deeper feelings. Not only do you fall back on other—usually irrelevant—factors when you analyze rationally, but you also begin to apply odd theories that you would never use without thinking.

Suppose your favorite fruit in the rich fruit bowl at home that is bursting with vitamins and variety is oranges. Now I want you to draw up a list of the fruits that you are going to take to work in the next 5 days for eating during your lunch break. Despite the fact that oranges are your favorite fruit, it is unlikely that oranges will feature five times on your list. Research has shown that this is something we rarely do. We tend to draw up a varied list along the lines of two oranges, two bananas, and a pear because a funny theory in our head tells us we crave variety in what we eat and drink. And our final decision is based on that flimsy theory. But what do we do in real life? Just looking at myself, for years I have been taking every day that piece of fruit that I like best: oranges.

What is happening here? Variety in food and drink is important to us, but only in the short term. I do not want to eat five oranges a day, but I cannot turn down one or two a day. Our problem is not that we fall back on such funny theories when we actually decide what we are going to do but that we use them more often when we start to reflect on choices to be made. The more that we deliberate on intuitive judgments, the more that misleading arguments and all sorts of rationalizations come into play that eventually might well screw up everything. Conscious thoughts can impair the process of decision-making.

Did the list of fruits not convince you? Perhaps the research done by S. Iyengar and R. Fisman on speed dating might. Participants of a speed date were asked to fill out a short questionnaire on four occasions: just before and immediately after the speed-dating evening and 1 and 6 months after the event. They were asked to rate on a scale of 1 to 10 the personality traits that they were looking for in a prospective dating partner. Some of those features were appearance, shared interests, sense of humor, sincerity, intelligence, and ambition. And when the big moment had arrived, they were asked to fill out the same questionnaire again after each encounter, but this time they had to rate the person they had just met.

A remarkable pattern emerged across the different participants and the various questionnaires. Individuals who had given the highest score to intelligence and integrity before the start of the speed date gave the most points to completely different categories after the event, such as sense of humor and appearance. In the lists where the speed daters had rated each other after each encounter, the scientists found the answer to this remarkable and sudden change in preference. The personality traits of the person they felt most attracted to during the speed date event came to the surface, and this had made them change their mind about the kind of person they were looking for. As quickly as that change had surfaced, it was gone again. One month later, exactly the same features as before the night of the speed dating scored highest again.

Before the evening and 1 month after the speed date, the participants were daydreaming and verbalizing about their Prince or Princess Charming; it was then that they assigned weights to the personality traits of a person that were different from those during the big night itself, when they chose purely on feeling or intuition. A criterion that weighs more when verbalizing is, for example, how much their parents would like the prospective mate. Before the start of the speed date, they found it important to come home with a partner that their parents would find suitable as a future son-in-law or daughter-in-law, but when Cupid shot his arrow, they fell for that one true love and the only argument they could come up with to justify their choice was nothing more than "it just clicks." It all boils down to a gut feeling you cannot explain. The most important decision of your life is largely taken on the basis of feeling or intuition.

In 2021, in total 12,031 marriages and about 4,900 legal cohabitations were dissolved by way of divorce in the Dutch-speaking northern part of Belgium.[10] Intuition undeniably exerts an influential force on our thinking, but it is clearly not infallible. It can be distracted, suppressed, or overridden by other conscious or unconscious workings of the human mind.

Can a better knowledge of decision-making help us make better choices? Modern life is filled with puzzling choices. We not only have to choose more often but can also choose among an increasing number of alternatives. You would think that this wide array of alternatives would have brought us only benefits because it gives you and me more freedom of choice and enables us to find exactly what our heart desires. However, the opposite is true: too many alternatives lead to indecision. They affect one another's attractiveness in a negative way, so that we are no longer inclined to choose at all.

I know that feeling all too well. In the local supermarket I am struck by it each time when fabric softener or washing powder features on the weekly shopping list and my wife has forgotten to show me the exact product she wants. How can I pick the right product in that 20-meter-long laundry detergent aisle stacked five shelves high with washing powders in all possible colors and scents? I have repeatedly come across men who at that moment, instead of seizing the opportunity and choosing themselves, call a helpline. To get rid of the choice stress as soon as possible, we reach for the smartphone in our back pocket and call home for further instructions.

There are roughly three ways of choosing:

1. The quick choice where you hardly think or do not think at all
2. The unconscious choice where you sleep on the problem
3. The rational choice in which you analyze the alternatives and carefully weigh the pros and cons

For almost everyone, the first way of choosing is the worst manner, the second is a bit better, and the third is downright the best. But is this correct?

From the various examples I have already given, you know better. In fact, all three manners are equally good, provided you use them in the right place and—even more important—at the right time in the decision-making process. Different situations require different modes of thought. Simple problems that can be solved very precisely with a limited set of strict rules (e.g., mental arithmetic or buying clothes) are best tackled with your working memory or conscious mind (option 3). When dealing with a complex problem (e.g., accepting a job offer or buying a house), you better let your unconscious mull a night over the many different relevant features and pieces of information that have to be combined and integrated (option 2).

Why? Our conscious working memory with its processing capacity of 60 bits per second suffices to solve problems where you only need to use a limited amount of information. As soon as the amount of information that has to be processed increases, the quality of conscious thinking drops rapidly. Your conscious mind simply becomes overwhelmed when you exceed its memory-storage capacity. Then it is time to switch to your unconscious mind that, with its much larger processing capacity, can deal with complexity far better. You might be fast asleep, but your unconscious never sleeps and keeps working on the problem. Precisely for this reason unconscious thinkers make by far better decisions in complex problems after a good night's sleep or a brief period of relaxation than conscious thinkers.[11]

And what about the quick choice or the intuitive judgment (option 1)? Within your field of expertise you can make accurate and precise judgments within a fraction of a second. Prolonged practice and a great deal of experience counts for something. That is exactly what the German Egyptologist Erman did. A quick glance at the hieroglyphic text inscribed on the commemorative scarab of Nekao allowed him to expose the artwork as a blatant forgery. As he is browsing through the numerous card-index boxes of the *Wörterbuch der aegyptischen Sprache*, his face brightens. He was not mistaken; his intuitive judgment was correct. All kinds of things are wrong with the inscription. The text on the scarabs turns out to be a pastiche of three Egyptian inscriptions published only a few years earlier. According to Erman, the forger could read simple hieroglyphic texts. He had a superficial basic knowledge of the language, but he clearly did not fully master all the intricacies of Egyptian grammar. Immediately Erman publishes his findings in a leading international academic journal, and 1 month later, at an international congress held in Berlin, he triumphantly lets the most renowned Egyptologists of the time, Capart included, sign a statement wherein it is stated unequivocally that the two scarabs are forgeries.

Where did the expert intuition of Capart go astray? The seller of the scarabs, the son of a well-respected French Egyptologist, assured him that the artifacts had been excavated by his father in the Nile delta: provenance was assured. Reputed Egyptologists assured him that the Egyptian text was not a forgery: authenticity of the text was assured. The more Capart verbalized the whole matter, the more convincing all scientific data appeared to be in his favor: the scarabs could only be authentic. Then he got so thrilled by the idea of acquiring archaeological artifacts confirming the circumnavigation of Africa by the ancient Egyptians for the museum in Brussels, that he wished both masterpieces were genuine. He held his wishes as truth, after which he could not be fast enough in Paris to buy both commemorative scarabs.

Monday,August 19, 2019

Without breakfast I rush to work. I have to call my colleagues at Aarschot. I left the Aarschot fire service 7 years ago, but I still have the phone numbers of my former colleagues. They are on a list in one of the many folders in my office at the fire station.

On Friday evening, August 16, numerous firefighters have come to the fire station of Heusden-Zolder to pay their last respects to the regretted colleagues. When I leave the fire station, I pass by the long line of firefighters who are patiently waiting their turn. A young man addresses me. His face looks familiar, but I cannot place him. I start to look around, and then I recognize the people standing next to him. My colleagues from Aarschot are here! I burst into tears, embrace one of them, and cry on his shoulder. Between the sobs I tell him my story. That very moment the battalion chief of Heusden-Zolder comes to fetch me. He leads me to the parking lot of the fire station, where all personnel of our Fire and Rescue Service are brought together after they have shown their last respects. What a pity. For me the moment with my colleagues of Aarschot might have lasted longer.

The first colleague of Aarschot I call that Monday morning tells me that he almost stood on my doorstep last weekend. For his work he had to be in Beringen, but dropping by unannounced seemed inappropriate. A missed opportunity, but we are going to make up for it on Friday evening, August 30, in the city of Aarschot. Our talk ends abruptly. His pager starts to beep. He has to leave. I urge him to be careful.

I call the next colleague. We reminisce about an incident he has long forgotten. A storm had raged over Aarschot and the only tree still standing in a courtyard threatened to fall on a village hall. It never bodes well when firefighters already on scene request the assistance of an officer. That night was no exception to the rule. I summoned the colleague to come to the courtyard for consultation. In my team nobody knew more about felling dangerous trees than he did. The task could only be brought to a good end if we somehow could let the huge tree fall in the opposite direction. There was just one problem: None of the low tree stumps in the enclosed courtyard could be used as an anchor point for our Tirfor winch. “Is the winch on our old Unimog perhaps an option?” I asked him. “If the vehicle gets through the narrow entrance gate and the steel wire on the coil is long enough, it might do the trick,” he replied. I asked the dispatcher to send the Unimog. The driver was confident: "It is possible, albeit a close call." And whenever something can be done, the fire service does it.

The last firefighter of Aarschot I call is from the same class. We both entered the fire service on the first of April 2007. He does not pick up his phone. I leave a message on his voicemail. In the afternoon I try again. This time he answers the phone. He is in Holland, on board one of the Gondeletta boats of the Efteling theme park. We keep it short.

I am really looking forward to next Friday! I cannot wait to see my former colleagues again in the fire station of Aarschot.

6. AM I GOING MAD?

Today is Tuesday, August 20, 2019. We are 9 days after the fatal incident. That day and the previous ones I could link everything I was seeing and thinking without effort to the tragic fire, as if my whole life had been predestined for just that one incident. I felt like I was in the movie *A Beautiful Mind*. This movie from 2001 deals with the life of the Nobel Prize–winning mathematician John F. Nash, who suffered from paranoid schizophrenia. A major difference is, however, that everything I observed was not a figment of my imagination but also there in the real world. It was not hallucinations or flashbacks but sensory cues in my immediate surroundings that triggered weird associations.

Should I have been worried about those strange sightings and psychotic experiences? Not really, but back then I did not know that traumatized people have the tendency to superimpose their trauma on everything around them. The thread perception system inside their brain has changed. Furthermore, large-scale studies indicate that no less than 10%–20% of the population has short-term psychotic experiences (hallucinations and delusions) without much harm. According to the Belgian biologist and science philosopher G. Magiels,

> It is now clear that psychotic experiences occur in the general public without always leading to problems. As we are all carriers of the same genes, in different variants and frequencies, that is not surprising. It is the combination of all those genes in interaction with the environment that determines who somebody is and how mentally sound he is or remains.[12]

And the environment I was exposed to earlier that month was somewhat traumatizing to say the least.

All the bizarre associations or psychotic experiences discussed in this section fall under what is referred to in the medical literature as *intrusive symptoms*. Such symptoms can manifest for 1 or more months after a traumatic experience. Occasionally, these distressing recollections linger for the rest of one's life. This category of symptoms includes not only thoughts or recollections of the trauma itself but also any object, sound, or image that evokes an association with the traumatic event. Take it from me: when these associations are suddenly and unexpectedly triggered, they always strike deep.

First, an example of such a strange association in the shape of a thought that is almost inevitable for every firefighter in Belgium. The patron saint of miners is Saint Barbara of Nicomedia, the current Turkish city of Izmit, some 150 kilometers east of Istanbul. Several professions dealing with unforeseen death invoke her intercession. The Belgian fire service is one of them. That night, two firefighters were pulled out of the burning building. Both found a safe haven on . . . Saint Barbara Street. In the coal-mining area of the Belgian province of Limburg, a Saint Barbara Street is unavoidable in mining municipalities like Beringen, Heusden-Zolder, or Genk. Of course there is no causal link between Saint Barbara Street and the rescued firefighters. Someone who does not work for the fire service would never make this link that is so obvious for every Belgian firefighter.

The working principle of unconscious associations or intrusive symptoms is that all links from one's sociocultural environment come into play, including noncausal ones. The hyperactive propensity of our brain for causal relationships (cause and effect) is so outspoken that we even infer connections or perceive patterns where there is actually little more than a random process. Delusions are this trait in the superlative.

The associations I discuss next will come across a lot stranger and less coherent than that of Saint Barbara Street. They are not collective but individual associations because the ordeal that I endured that night has reorganized the way my mind manages perceptions. They will seem bizarre, irrational, far-fetched, and even crazy.

Tuesday, August 20, starts nice. I have a day off, and I have planned a bike ride with my parents and two sons. We do that every summer holiday. This time we are going to cycle through the trees in Hechtel-Eksel. This unique cycling experience in the green heart of the province of Limburg lets you cycle around in a double circle all the way up to the treetops, up to 10 meters high. After breakfast we get on our bikes and cycle to the city of Leopoldsburg. A truck crosses our path near the defense training estate.

Nobody notices anything special, but I am instantly catapulted back to the fireground. The cargo bay is one large billboard. It depicts two firefighters in protective clothing and with fire helmets on standing in an immense sea of flames. One of them—coincidentally or not, a sultry and seductive woman—holds a fire extinguisher in her hands. With the accompanying slogan "Call us before you get hot," the company Somati tries to win new customers for their fire extinguishers. This is just plain advertising, but it has a totally different impact on you when under your command two colleagues have perished in a sea of flames like you have never seen before. I must take a picture of the truck. I leave the group and pursue that truck. Fortunately for me, 300 meters further down the road it stops to make a delivery. I can count on the driver's understanding, and with pleasure he parks the truck out of the shadows of the unloading dock so that I can take as many photos as I want in good light.

I get back on my bike. The boys and my parents have stopped and waited for me. We are back together in no time. A few hundred meters further we pass by the yacht harbor of Leopoldsburg. On dry land stands a pleasure boat that is half burned out. Immediately I am back again at the incident ground. This time I am explaining the course of the fatal fire to a colleague fire officer of the Fire and Rescue Service of North-Limburg. What is the link between the boat and the fire? A few weeks before the fire, I attended a meeting at the fire academy of the province of Limburg. My colleague and training coordinator from North-Limburg had bandages around his fingers. I asked him what had happened to his hand. He told me about an incident with a pleasure boat that had caught fire in a lock. The boat had been drifting toward other vessels, but by grabbing a handrail, he managed to stop it and avoid a collision. However, such a handrail of metal conducts the heat very well. Without protective gloves you will burn your fingers. As I cycle past the boat, his story flashes through my mind. That colleague with the bandages is the fellow fire officer that I first ignored that night but to whom I later told my story for the first time. I cannot resist sending him the pictures I took of the vessel via WhatsApp. Many months later we come back to that message. Fair-minded as he is, he tells me that he found my message bizarre and had no idea what it meant. That is typical for such associations.

For me everything is obvious and clear. With the greatest of ease I can link just about everything to the fire as my thinking has become more fluid and is no longer constrained by the rules of logic. However, for the people with whom I share my racing thoughts, they are puzzling.

Back on the bike again, the life of Saint Norbert flashes into my mind. Do not ask me where it comes from, but all of a sudden, the idea of posttraumatic growth in the Middle Ages occurrs to me out of the blue. I explore this epiphany in more detail in the next section of this chapter.

Creativity is not the result of hard work done by the conscious mind over a long period of time but the result of the continuous work done in the background by the unconscious. While you are busy going about other things, your unconscious mind continues to mull over important problems that have not yet been solved or decisions still to be made, and then out of the blue, it comes up with a new way of understanding the problem and delivers that novel solution to the conscious mind. This process is never straight to the point but more a long and winding road. Hence the difference in quality of the associations and thoughts at play in my mind. Some are for sure far from my best ideas and definitely unfit for publication, while others with some additional research turn out to be not nearly as bad.

Search processes by the conscious mind have a narrow focus and are very goal oriented. Those by the unconscious, on the other hand, are very associative and fan out strongly with no real focus. Compare it to fishing: instead of a fishing rod the unconscious uses a dragnet. A richer catch of unusual and original solutions is guaranteed. Conscious thinking can never achieve that result because it is a lot less creative than the unconscious. The novel insight is offered to your conscious mind when there is processing capacity available, when it is doing nothing in particular and you are just daydreaming without really focusing on anything. Many have that *eureka* or *aha* moment when they just sit with a blank stare in one of the three *b*'s: bus, bed, or bath. As for me, it always hits me while I am running.

In the days after the fire, I started to run more than usual, for a double reason. I hoped it would clear my head and tire me out so I would sleep better and longer than in those miserable blocks of 2 hours. My normal sleep-wake rhythm was completely gone. With regard to clearing my mind, exercising has the exact opposite effect. For some reason I can think better while walking or running. The ideas then flow freely at the tempo of my normal tranquil pace. This for sure was the case during my run on Thursday, August 15. In a moment of clarity I realize there is a gap in the current model of reading the fire: an incident commander always reads the fire under the influence of the incident type that they have chosen intuitively immediately upon arrival at the fireground. It is no coincidence that just after the fire, not only so many bizarre but also creative ideas just pop into my head. Creativity and madness have a common genetic basis. Biologist and science philosopher G. Magiels describes it this way:

> Nevertheless, everything indicates that there is a deep genetic connection between the creativity of people with psychosis (which can sometimes become excessive) and those who push boundaries in science or art (by seeing connections that others have not seen before).[13]

Now back to Tuesday, August 20. In the late afternoon I am back home after a nice bike ride, which abounded with psychotic experiences. In the evening Dad's taxi service is out on the road again. Jonathan has to be taken to Dessel for a football game. It will be one of our most memorable car rides ever. When he looks back on my monologue, he still has to laugh about it. With neophyte zeal, I try to convince my eldest son—not yet 13 at the time—of my novel insight into posttraumatic growth in the Middle Ages. Then my smartphone rings. Immediately his face brightens. I can see him think: "Saved by the bell!" He welcomes anything that will stop my crazy ramblings about a medieval saint. We cannot call hands-free in our car, so I ask him to answer the phone. My psychologist is calling. She wants to fix a date for our next appointment. I say I will call back in a minute because we are almost at our destination. A few minutes later, she calls again. When I inform her that bizarre associations are pouring in all the time, she is concerned about my mental health. Now I understand why: there is a thin line between intrusive symptoms and delusions. A new appointment must be arranged soon and preferably the next day.

After the phone call, I am furious with her. Why is that psychologist talking nonsense! In passing she let slip that I do not show any emotions, while the last few days I have been experiencing myself as a vessel of emotions that is literally overflowing. On Friday evening, August 16, I cried on the shoulders of one of my former colleagues from the Aarschot fire service after having paid my last respects to the deceased firefighters in the fire station of Heusden-Zolder. A day later at the funeral, the song "No Sound but the Wind" of the Editors starts. I crash on the lyrics: "Help me to carry the fire. We will keep it alight together!" This is obviously a metaphor, but in my state of mind I take it literally. I am back at the fireground and the last thing I want to keep alight is the fire in which my colleagues are trapped. Over the past few days I have done nothing but hug just about everything and everyone. Fellow firefighters or the lady who organized the silent walk that took place in Beringen on Sunday, August 18, it is all the same to me.

After the phone call, my brain immediately stops making those weird associations as if the psychologist has become the mistress of my wandering mind. Or at least for a short while, because when I stand before the entrance of her practice the next morning at 8:30 a.m., I find myself again at the fireground. The dimensions of the outside door are unusual. The door is high, but much narrower than a regular door. In my mind, it is the same size as the opening at the back of the former Delhaize in Beringen-Mine, through which I have sent in my colleagues of Heusden-Zolder that night . . .

Where do all those bizarre associations come from? They all come out of my context. Out of my own reality. For intrusive symptoms, like for confirmation bias and psychosis, context is everything. As always, more than 90% of that context is determined by top-down mental processes operating below the surface of awareness. Already at the fireground, in my endless wandering around the building, I began to search for answers. I had to figure out what I had done wrong. However, I was not the least bit aware that my subconscious would keep on ruminating over that problem over the following days and weeks, while I was trying to pick up the pieces of my life and move on to something else.

The various experiences just described and serious psychiatric conditions might conceivably have a common basis. According to the *continuity hypothesis*, which is currently on the rise in psychiatry, a great variety of disorders show such a strong resemblance that they must have one or more underlying factors in common. There is indeed more overlap than difference. Serious psychiatric disorders arise from psychological functions that we all have. They are in essence the same as the experiences that you and I have every day but in an extreme form. An unhealthy paranoia results from suspicious thoughts gone awry. The only points of difference are quantitative: *how much* and *how long*. As with accidents, nobody is exempt from such disorders of excess. A mild or even a severe mental health condition can befall anyone.

During a psychosis, the production of the neurotransmitter dopamine surges in a certain brain area (the striatum), while in another part (the frontal cerebral cortex) it is greatly lowered. With the chemical messengers in our brain, everything boils down to balance. When distorted, the delicate electrochemical balance between inhibition and stimulation of thoughts, feelings, and actions that you need to function normally is disrupted. You can become overstimulated or understimulated.

Dopamine is best known for the role it plays in the mesolimbic reward circuit of the brain. You simply cannot resist the delicious food served at breakfast or dinner because of the squirts of dopamine that are released inside your brain in anticipation of the pleasure that you are about to experience. That dopamine rush is also the reason why you desire to eat that extra donut or chocolate chip cookie even when you are no longer hungry. Not everything this chemical makes you pursue is necessarily better. The objects of your desire only appear to be better at the moment of your most intense craving. All in all, the whole circuitry is just one of Mother Nature's smart tricks to seduce you into doing the same thing over and over again all your life to ensure the survival of your organism and the continuation of the DNA stored in your unique mix of genes. Without dopamine, we would simply not bother to eat, exercise, or even have sex.

The Belgian pharmacologist Paul Janssen of Janssen Pharmaceutica or the current Johnson & Johnson is at the heart of the use of dopamine blockers in antipsychotic drugs. Around 1958, he saw many cyclists crossing the finish line with the symptoms of schizophrenia, more precisely auditory or visual hallucinations. Doping in cycling is of all times. In those days cyclists rode on amphetamines, or speed, as this chemical is now called in nightlife. The performance-enhancing effects of amphetamines were well known in the late 1960s. During World War II, many experiments were conducted on soldiers, not only by the Germans but also by the Allied forces (of course, without permission or knowledge of the soldiers themselves). On amphetamines, soldiers march and keep on marching, even as far as Moscow when they are told to! In addition to the feeling of intense concentration, amphetamines counteract physical fatigue, but they can evoke hallucinations and delusions. Pharmacologist Janssen began searching for a molecule that would block the receptors that amphetamine binds to. The race to develop dopamine blockers was on.

Dopamine also performs other functions in our brain. One of them is marking neutral events and their mental representation as special and indicating whether they are pleasant or annoying. Just as you highlight key words in a text with a fluorescent marker, dopamine marks certain sensory stimuli of the abundance of impulses that enter through the sense organs as important. Subsequently you pay attention to what is marked as important and ignore the rest.

But when dopamine runs wild and attaches salience to ordinary things, you are overwhelmed by that excess and become less and less able to separate the essential from the nonessential. Things that are of no importance whatsoever will be perceived as most pertinent. From here to schizophrenia (complete with psychosis and delusions) is only a small step. Various coincidences and random events take on a special meaning. Everything you notice—such as publicity on a truck or an ordinary combination of letters or digits on the license plate of the car in front of you—is a message or an indication of something special, and before you realize it, you are firmly convinced of ideas that do not make sense, relate everything to yourself, and think you are predestined for a special task. All this is bound to result in a whole series of strange behaviors.

When I enter the practice of my psychologist for the last time in the beginning of October 2019, the outer door with its unusual dimensions makes a completely different impression on me. Now I smile at the weird association or intrusive symptom it conjured up a few weeks earlier. With time everything starts to fall into place again.

Monday, September 9, 2019

I am hanging on the telephone again. This time I am calling former colleagues from the time before I joined the fire service. A headline in the news last Friday has initiated a new stream of telephone calls: My previous employer is laying off 1,400 members of staff. It is not easy to get hold of the head of the finance department I worked in. His wife answers the phone and thinks I am a telemarketer; after I have introduced myself and asked to speak to her husband, she tells me that they are not interested in a new electricity contract.

I am starting to go back further and further into my past, to the period before the fire, when life still seemed to smile at me. In an email a former colleague asks me how things are going. I reply:

> Now things are not going so well. Immediately after the fire I was living in a bubble and nothing could trickle in. I was continuously struggling with my own feeling of guilt and trying to support my colleagues from the fire services of Beringen and Tessenderlo, and from the fire academy. Now my adrenaline level is finally dropping. In itself, that is not a bad thing, but this gives unpleasant thoughts plenty of opportunity to enter my mind, and I am getting more than my fair share. More and more details of the incident come to light. And all that new information does not help me to maintain my peace of mind.
>
> I also notice changes in my behavior and my body is behaving strangely. When I have an idea, I have to act on it immediately and not five minutes later or I become extremely restless. Then I feel not only my pulse quicken, but also my temperature rise (up to sweat on my back!). My temper also has not improved. At the slightest provocation I become angry and I blow up.
>
> According to my psychologist—with whom I've also been furious by the way, luckily not in her presence, but I have told her—that's normal. She says that cooling down will take at least three months.
>
> We will see what my brain and body have in store for me in the coming months . . .

7. POSTTRAUMATIC GROWTH IN THE MIDDLE AGES

There is nothing new under the sun. Posttraumatic stress disorder (PTSD) surely is not new. Only the name is fairly recent. In the U.S. Civil War (1861–65), this psychotrauma-related stress disorder was called *soldier's heart.* During World War I (1914–18), it was known as *shell shock*, and after the war it was rebaptized as *war neurosis.* With the return of the countless traumatized soldiers of the Vietnam War (1955–75), the term posttraumatic stress disorder came into use. Now, nearly half of the U.S. combat veterans returning from the wars in Iraq and Afghanistan have been diagnosed with PTSD and a mild brain trauma (blast-induced traumatic brain injury).

In modern-day psychiatry, a distinction is made between an acute stress disorder and a posttraumatic stress disorder. These disorders are a lot more common than you would imagine. They are a fact of life. About 60% of men and 50% of women experience a psychologically traumatic event during their lifetime (e.g., accident, physical attack, domestic violence, rape, childhood sexual abuse, natural disaster, war). An acute stress disorder lasts less than 1 month. Both disorders occur after events that trigger acute anxiety, but PTSD happens in a context where there is usually a threat to the victim's physical integrity (e.g., mortal danger, death threat, sexual assault). In 10%–15% of the cases, the fear gets so thoroughly ingrained in the brain that the person struggles continuously with the painful aftermath of the traumatic event and has intrusive thoughts, flashbacks, or nightmares for the rest of their life.[14]

And what is posttraumatic growth? To put it in the words of the American firefighter J. Knapp: "A normal reaction to an abnormal incident."[15] Any traumatizing life event will strike deep. The negative side is quite obvious and painful, but luckily there is also another side. After the dust has settled, you as a victim can emerge stronger from the horrible experience if you succeed in putting the traumatic event behind you, withdraw attention from it, and think less and less about it. In my case, the focus shifted from the fateful fire to a quest for answers: How could this and other fires have ended so dramatically? What is the anatomy of those fatal fires? The result of my posttraumatic growth is the book you are reading now.

Over the past centuries, posttraumatic growth has led to many great things. In my opinion it is at the heart of more than one monastic order. In the autumn of 1986, I went on a retreat in the Norbertine monastery of Tongerlo together with the rest of the fifth-year students of my high school. What do you get when you unleash a group of teenagers with only two supervisors for a few days and nights in a monastery? Wild nights and, in the morning, many sleepy eyes around the meters-long breakfast table in the imposing refectory of the abbey. Fortunately, other things have also made a lasting impression upon me.

First and foremost, there is the copy of *The Last Supper* of Leonardo da Vinci, which belongs to the select group of oldest and most faithful copies of the original fresco and on which, according to the American art historian J.-P. Isbouts, the grand master himself most likely also did some painting. In 1545, the abbot purchased the impressive canvas of 34 m^2 in Antwerp and gave it a prominent place in the choir of his brand new abbey church. The painting, made around 1507 mainly by pupils or followers of da Vinci, might well have been commissioned by the French king Louis XII. He is said to have admired the original mural in the refectory of the Dominican convent of Santa Maria delle Grazie in Milan, Italy, so much that he even considered having it carved out and brought to France. Only after that plan proved impracticable did he settle for the next best thing: a full-scale copy on canvas.

I also remember the life of Saint Norbert, which one of the monks told us about during a guided tour of the abbey. Norbert, born around 1080 in Xanten, Germany, belonged to the nobility of Lower Rhineland. In accordance with good feudal custom, his elder brother inherited the entire family estate. Norbert, on the other hand, got an ecclesiastical benefice as a consolation prize, but unquestionably more important to him was the revenue he gained from the endowment attached to that church office. Norbert was not called by God to serve in the Church, which was quite obvious from his way of life. He may have held ecclesiastical titles like *choirmaster* and *canon of the Saint Viktor's Chapter at Xanten,* but he does not exactly stand out because of his pious conduct. On the contrary: If you were looking for a great party or wanted to spice up the party you were about to throw, then you had to turn to Norbert. He was always showing off with the revenue from his benefice, and with grandeur and panache he knew his way around the court of Henry V, the emperor of the Holy Roman Empire. But this debauched life ended abruptly in 1115.

One sunny day in spring, Norbert, then 35, was riding his horse through an open field in the company of a friend. He was going to see his sweetheart in Vreden. Other sources, however, claim that he was set on paying a visit to the ladies of pleasure. All of a sudden the skies darkened and a lightning bolt penetrated the clouds, striking just in front of his horse's feet. The steed reared and threw him. Norbert got knocked out, and when he regained consciousness 1 hour later, he asked God in a prayer what he had to do. Sometime later he repented (notice: a *canon* repents!) and he was ordained deacon and priest.

Already as a 16-year-old, I was intrigued by this strange story. By falling off his horse he must have lost his marbles. Hence, he transformed from a party animal into a hermit who exchanged all the glitter and glamour of the imperial court for a life of strict penances. The fatal fire of Beringen altered this rather simplistic view of mine, for now I myself was in troubled waters because of an acute stress disorder.

When I go to sleep that Sunday evening, August 11, and close my eyes, geometric figures in the shape of stars and deer gambol before my eyes. The deer closely resemble the lighted reindeer decorations that pop up in people's front yards at Christmastime. When I open my eyes, everything has vanished. The effect can best be compared to a bad trip on LSD, although I am not an experienced expert when it comes to LSD. The next night is even worse. As if struck by lightning, I wake up in the middle of a nightmare. I am wide awake, just like when my pager beeps in the middle of the night. I dreamed that I was back at the fireground, but this time I was trapped inside the building and there were flames everywhere. Fortunately, it stayed with just that one nightmare. But for a long time, getting through a night would remain much harder than passing a day.

My restlessness is extreme the first days and weeks following the fire. When I have an idea, I have a compulsive urge to act on it immediately. Furthermore, I have changed into a motormouth. Talking is what I do best, and once started, I simply cannot stop. I feel the urge to express my emotions and talk to everyone about the incident. There has never been a period in my life wherein I have spent more time on the phone. A phone call of 3 hours is my personal record! This is probably also true for the unfortunate colleague on the other end of the line. At some point I get muscle pain in my upper right arm. Did I make a wrong move or bump into something with my right arm while exercising? I struggle to recall something like that, but then it dawns on me. Holding the phone to my ear several days in a row has begun to weigh on the muscles in my upper arm.

Another striking feature is that I easily get tired. My brain is way too preoccupied with the incident. My mind is racing, not one second of rest. That costs energy, a lot of energy. Add to this bad nights where you awaken every 2 hours, and at the end of the day you are tired, dog-tired. Several evenings I go to sleep before my sons and with a headache, for I constantly have a nagging headache. Four weeks after the incident, the adrenaline levels finally begin to lower.

The first weeks after the incident I lived on an adrenaline rush. That did have one advantage: horrible thoughts did not get the chance to enter my mind. That starts to shift after a month. The stress is, however, clearly not completely out of my body yet because I still frequently get headaches, I am short of breath, and my temperature regularly rises higher than normal. The motormouth finally starts to sputter. The sound of silence is good not only for me but also for those around me.

I just cannot let go of the idea about posttraumatic growth in the lives of medieval saints. It flashed into my mind a few days ago during a bicycle ride. I do not have a choice; I have to jump right on it. Through my training as an Egyptologist, I am familiar with the lives of early Christian saints from the ancient Near East. They are not like that of Saint Norbert or the various saints collected in the *Legenda Aurea* (*Golden Legend*) by the Dominican Jacobus de Voragine around 1260. In my opinion they can only differ. Different times have different manners. This results in another attitude toward religion in general and the Christian faith in particular. In the words of the historian Y. N. Harari: "Different socio-economic realities and daily routines nurtured different states of consciousness."[16]

In the early Christian period, saints were regarded as pioneers. Before Christianity became state religion, there were several periods in the Roman Era where Christians fell prey to violent religious persecution. One of those victims is the Austrian Saint Florian, the patron saint of firefighters, who in the third century reportedly organized and trained an elite group of Roman soldiers whose sole duty was to fight fires. When this retired army commander refused to offer sacrifice to the Roman gods in 304, he was thrown in the Enns River, a tributary of the Danube, with a millstone tied around his neck. The day of his execution, May 4, is nowadays known as the International Day of the Fire Service.

In Egypt, the persecution under Emperor Diocletian was the most bloody. For this reason, the Egyptian Christians or the Coptic Church have chosen the first year of his rule—284—as the beginning of their era. For several early Christians, the religious persecution was the ideal opportunity to actively seek martyrdom. They began to defy the Roman authorities ostentatiously. Until the fourth century, martyrdom offered in the eyes of most Christians a guarantee of direct access to an afterlife with the saints, the early Christian variant of the 72 virgins that the suicide bombers of the Islamic State believe they will be rewarded with in paradise.

From the second half of the third century to the sixth century, tens of thousands of devout Christians in Egypt, Palestine, and Syria exchanged the material comforts, worldly politics, and secular social distractions of urban life for the forbidding solitude of the scorching desert. In quest of purity of heart and soul, they renounced all things of this world and dedicated themselves to prayer and rigorous asceticism. One of the most famous anchorites or hermits is Saint Anthony of Egypt (251–356), who found shelter in one of the countless tombs that, during the centuries-long rule of the pharaohs, were carved out of the bedrock of the Egyptian desert all along the banks of the river Nile. From an iconographic point of view, this bearded saint in brown cloak is easy to recognize by the pig at his feet and the T-shaped staff with a bell in his hand. The hallucinations to which he fell prey in the desert while resisting his urges for water, food, fame, status, and sex are aptly depicted by the Belgian painter Félicien Rops in his work *La tentation de saint Antoine* (*The Temptation of Saint Anthony*). In this drawing from 1878, none other than Satan himself replaces the body of Christ on the cross with a voluptuous naked lady, all before the eyes of an Anthony driven to despair.

In the absence of input, brain circuits start to generate information by themselves top-down and interpret that input as if it had entered bottom-up through the sense organs. Contemporary fears and preoccupations surface in the shape of hallucinations. Malnourished and in desolate solitude, Saint Anthony and many other fathers and mothers of the desert had to deal with exactly what they had previously imagined and stored in their mind. If you are keen on having your firmness in faith tested by the devil, then Satan is bound to knock on the door of your humble abode in the middle of the desert under a sweltering sun. The lives of the desert fathers and mothers as well as their sayings are bursting with temptations by the devil and the vices of the flesh, which they consciously sought to overcome with their uninterrupted prayer and great physical mortification.

In these times of progressive secularization, Satan and angels are on the decline in hallucinations and delusions. Thanks to Hollywood, themes like secret police, espionage, and terrorism prevail in Western culture. In Islamic countries that role is assigned to *jinns* or spirits, and in African cultures, to voodoo practices. Hallucinations and delusions are firmly rooted within our own culture. This insight has been known for quite some time in religious thinking. Already around 500 BC the Greek philosopher Xenophanes of Colophon denounced the projection of the sociocultural background together with human whims to the household of the gods established on Mount Olympus.

In the Middle Ages, the attitude toward Christianity was completely different from that of early Christian times. In those days everything in our region was imbued with Catholic faith and even more with superstition. As a consequence, people reacted differently when they were suddenly confronted with something that was beyond their control. The psychological experience was totally different, and more than likely the inevitable inner search for an explanation would lead to God.

One saint is thrown off his horse and subsequently founds a new monastic order; another one falls from the citadel of Pamplona, Spain, and establishes the "shock troops of the Vatican." That is what happened to Ignatius

Loyola, the founder of the Jesuits. This youngest son of a noble Basque family with 13 children, 4 of which were probably bastards, was fully immersed in courtly life before his conversion to Christianity. This jolly officer was "frisk in the game and in dealing with women, in quarrelling and wielding weapons. He was tempted and conquered by the lust of the flesh."[17] As a vain knight with a too-big ego, he was eager to gain fame on the battlefield. That was exactly what he was doing when the French troops shot him with a cannonball from the citadel of Pamplona on May 20, 1521. His right tibia was smashed in the fall, but during his slow and difficult recovery, in which his life had hung by a thread, his ego suffered a serious blow. While recovering from surgery, he began to read books on the life of Christ and the saints. This was not out of interest; there were simply no other books in the castle of his parents. Once recovered, he decided in February 1522 to radically alter his way of life. He devoted himself to God and began to wander around as a mendicant.

And then there is also Francis of Assisi, Italy. Before his conversion, his way of life was also not a model of Christian virtue. This self-assured wild child let the money of his father, a wealthy cloth merchant, flow. He was friendly to deal with and excelled in just about any sport. Always well-dressed according to the fashion of the day, this handsome young man did pretty well with the ladies. His dream was to become a knight. Yet he is a question mark because his conversion took place in various stages. Phase one contained several elements that can be described as traumatic. The reality on the battlefield was (too) far away from the courtly ideal of chivalry, for in 1202 at the age of 20, he was taken prisoner of war after a battle. He spent 1 year in prison, and after his release he fell badly ill for a long period of time. To be on the safe side, I give him the benefit of the doubt.

What if the fall of Norbert and Ignatius had resulted in an acute stress disorder? What would that imply in the Middle Ages? Already 2 days after the fire, I had my first session with a psychologist.

In the Middle Ages, the medical knowledge of physicians, doctors, and barber-surgeons was so insignificant that it provided excellent opportunities for hypocritical quacks and boastful charlatans. With a little help from a member of the audience that gloomily feigned an illness, they staged healing spectacles in marketplaces and fairgrounds. The Black Death or Plague was thought to be the wrath of God, which people had inflicted upon themselves with their immoral behavior, greed, blasphemies, heresies, and all the rest that was considered to be sinful in the eyes of the Catholic Church. Medieval men and women were not familiar with bacteria or with good hygiene. Streets served as open sewers and gave flea-infested rats free rein, enabling all possible diseases to spread at superfast speed.

Mental health conditions like depression, epileptic seizures, and psychosis were caused by demons in the head. What better way to let those tormentors escape than by drilling a hole in the skull? Whether the stone of madness was cut out by a physician who was a member of the craft guild or a quack, the result of this placebo surgical operation was precisely the same: At best you remained schizophrenic, and at worst you became even more so. Pieter Bruegel the Elder could not help but laugh at the gullibility and folly of his fellow citizens in his engravings (or those ascribed to him) devoted to the stone operation.

There they stood then, Norbert and Ignatius. Like me, they might have fallen into a severe existential crisis. They began to question themselves, but even more important, their unconscious minds began to look for answers. Shortly after their traumatic experiences, their levels of hormones and neurotransmitters would have been far from normal. They might well have experienced intrusive symptoms and began to see in everything indications for a bigger scheme or for their destiny in life. Considering the worldview prevailing in those days, I am not surprised that both concluded that God must have spared their lives for a reason, an important task, or even a divine mission. Their brains could handle neither pure luck nor sheer bad luck. There must have been a reason for everything. There were no neurologists or psychologists to explain the mental processes that were upsetting their mindset. The course of Ignatius's life is striking in this respect. I have never read more voraciously in my entire life than after my traumatic experience. I could no longer keep up with the grey matter between my ears. In order to get a grip on the new reality I was facing, I started reading about neurology.

But on what could Ignatius have laid hands? With only books on the life of Jesus and the saints in the ancestral castle, he could have ended nowhere else than with God. Both Norbert and Ignatius could only conclude that God must have given them a severe warning that they should reflect seriously on their way of life. They welcomed that second chance and grabbed it with both hands. Their days of a wild and dissolute life were over as they repented and began to wander around as mendicants, proclaiming a radically evangelical way of life.

Is there a grain of historical or neurological truth in the conversion of Norbert and Ignatius? Or is it just a topos, or fixed theme, in medieval hagiographies? Deliberately exaggerating the way of life of a saint before his conversion and portraying him as unchristian and indecent as can be is the best way to emphasize the miracle of his conversion and to make clear that God's ways are inscrutable. The act of falling itself is a topos that can be traced back to the passage on the conversion of Saint Paul—or in Hebrew, Saul—in the Bible (Acts 9:3–9, NIV):

> As he neared Damascus on his journey, suddenly a light from heaven flashed around him. He fell to the ground and heard a voice say to him, "Saul, Saul, why do you persecute me?" "Who are you, Lord?" Saul asked. "I am Jesus, whom you are persecuting," he replied. "Now get up and go into the city, and you will be told what you must do." The men traveling with Saul stood there speechless; they heard the sound but did not see anyone. Saul got up from the ground, but when he opened his eyes he could see nothing. So they led him by the hand into Damascus. For three days he was blind, and did not eat or drink anything.

Before his conversion, Paul was known as a rabid persecutor of the followers of Jesus, but now Christ was addressing him directly. On the basis of this psychotic ecstasy and temporary blindness, the Dutch neurologist D. Swaab makes the presumptive diagnosis of temporal lobe epilepsy. The temporal lobe is located on the side of our brain, just above the ears. Temporal lobe seizures can consist of direct contact with God or other religious figures, such as the archangel Gabriel, and can last anywhere from 30 seconds to several minutes. In some cases, the seizure modifies the personality profoundly and permanently. Patients become hyperreligious and between the epileptic seizures display symptoms of *Geschwind syndrome*, which is characterized by, among other things, a lot of writing. Other historical figures who most likely suffered from this rare form of epilepsy include the prophet Muhammad, Joan of Arc, Vincent van Gogh, and Fjodor Dostoevsky. With pleasure I leave further research into a possible neurological grain of truth in the lives of medieval saints to the Bollandists, if these Belgian Jesuits have not already come to the same conclusion a long time ago.

Compared to Norbert and Ignatius, I am keeping it low-key. A book on the Beringen fire and perhaps two or three articles in professional magazines, there it ends for me. We live in different times. To start a new religious order at a time when one monastery after another has had to close its doors, that would be proof of little sense of reality. But any publication of mine fades into nothing compared to the abbeys of the Norbertines. To this day we reap the benefits of Norbert's posttraumatic growth. Enjoying delicious cheese from the Postel Abbey or eating an ice cream in the shadow of the Averbode Abbey: We owe all these subjective experiences to Norbert, who one sunny day was thrown off his horse. Is it not amazing what posttraumatic growth can lead to?

Tuesday, November 5, 2019

In one instant all the progress I have made during the past few months is swept away. After breakfast I am scrolling on the news app of my smartphone. A headline grabs me by the throat: "Two Firefighters Killed in an Explosion in Italy."

Overnight a powerful blast has occurred at an unoccupied farm about 60 kilometers east of the city of Turin. In addition to three injured first responders, two firefighters were killed. A third firefighter is still missing. The colleagues responded to reports of a blast, but shortly after their arrival, a second and stronger blast occurred.

I am stunned. With the message "without words" I share the bad news. Over the next few days, the death toll rises to three firefighters. The Italian colleagues did not stand a chance: The explosion was triggered on purpose. In the rubble, a gas canister is found equipped with a remote-controlled detonation timer. The insurance scam that the owner of the farmhouse thought he could get away with has claimed the lives of three firefighters. In what kind of world do we live?

After the fatal fire, countless sympathy cards arrive in the fire station of Beringen. The fire station of Heusden-Zolder has also been flooded with flowers and sympathy cards. All those cards were very comforting. Now it is my turn to send words of comfort to the colleagues of the Asti fire service.

Monday, December , is another tough day. The severe weather that hit the southwest of France has killed three rescue workers when their helicopter crashed near the city of Marseille. The pilot, copilot, and firefighter on board are killed. And then the worst has yet to come.

January 2020: Massive bushfires unprecedented in their extent and intensity have been raging for months in many parts of Australia. The plumes of black carbon have drifted across the Pacific Ocean and now affect the skies over cities as far away as central Chile. Nine firefighters are killed in the inferno, including three American colleagues who have rushed to the aid of our Australian colleagues. Their C-130 Hercules aerial water tanker crashed in the mountainous region about 400 kilometers southwest of Sydney.

The newspaper of Wednesday January 8 has pictures of the funeral service for one of our Australian colleagues. My heart is bleeding. His 19-month-old daughter Charlotte steals the show. She is sitting on the floor under her father's casket, eating from a bag of chips. She does not realize what has happened. And that is a good thing, for sooner or later the harsh reality will sink in.

8. BLACK LIVES MATTER

Conscious and unconscious thought processes, adrenaline, time pressure, regression, environmental factors, group conformity, perceptual narrowing, genetic makeup, experience (or lack thereof) . . . all I have discussed so far comes together with several other factors in a fatal shooting in the South Bronx (New York, US) in the early hours of February 4, 1999, in which the 23-year-old black street trader Amadou Diallo from Guinea was killed.[18] Diallo's senseless death is more than just symptomatic of today's America, where, with the regularity of the clock, persons with slightly darker skin have to deal with excessive police brutality.

"I can't breathe" were the last words of George Floyd, the 46-year-old black American who died on May 25, 2020, when during his arrest a white police officer pinned him to the floor by placing his knee between Floyd's head and neck for almost 10 minutes. The man could not breathe.

Unfortunately, this violence is universal and of all times. Steve Biko is just another victim. On September 12, 1977, lying on the floor of his prison cell in Pretoria, South Africa, this anti-apartheid fighter succumbed to the severe brain lesions resulting from numerous blows to the head sustained during countless, extremely violent interrogations conducted by one or more white police officers.

And then there are the excesses of our own colonial past in the heart of Africa. The horror in the Congo Free State, the personal property of the Belgian king Leopold II, is the setting of the 1899 novel *Heart of Darkness* by Joseph Conrad. The journey up the Congo River by the main character Marlow is an exploration of the human psyche that turns darker and darker as he steams upcountry. The takeover of the Congo Free State in 1908 by the Belgian government did not improve the fate of the indigenous population in the already heavily depopulated inland. The personal gain of a handful of Belgian investment companies that now held the reins of power prevailed again over the social interests and the well-being of the indigenous population.

During the interbellum, the overall appropriation of the vast mineral resources and the common exploitation of labor continued with undaunted zeal. The cost of the accelerated large-scale industrialization that the Belgian Congo was then undergoing was fully borne by the native population. Nowhere was this eminent utilitarian opportunism more apparent than in the colonial labor camps, which sprouted like mushrooms. The appalling mortality and morbidity are on par with those of the concentration camps created during the Anglo-Boer War in South Africa (1899–1902). According to conservative estimates, around 20,000 civilians died of hunger, contagious diseases, and deprivation in the work camps that the British had set up for the black population. The actual number is bound to be higher. Already about 200 BC humans were described as a wolf to a fellow man: *Lupus est homo homini*.

As for all our behavior, the starting point is the unconscious mind, more specifically our unconscious opinion about the fellow man with darker skin, which must be rooted deeply in the human psyche. The inadvertent stereotypes and biases that have penetrated the collective unconscious of Western society are negative. You may state hand on heart that you find discrimination abominable, but the sobering results of the Implicit Association Test that you can find in many books on psychology will tell a completely different story. Be sure to try it. It is not hard; you just have to go through a list of words and names while tapping one of the edges of the book with your left or right index finger.[19] Still not convinced? Search for "black doll white doll research." The hand of an innocent child does not lie. Even when you are convinced that you do not discriminate and you are completely opposed to it, you still do it without noticing.

Research by the American psychologist R. Fazio on white persons has shown that it is not their conscious opinion about black Americans that directs their behavior but their unconscious biases. Those negative group stereotypes and prejudices make not only their nonverbal behavior but everyone's more negative toward fellow human beings with a different skin color. Our subconscious is not always as subtle as we would like it to be.

This is the unconscious setting with which, shortly after midnight on February 4, 1999, four white police officers of the Street Crime Unit of the New York Police Department drive into Wheeler Avenue in the Soundview neighborhood. In the late 1990s, the drug trade was brisk in this poor and working-class area of the South Bronx. After a late meal and a talk with some of the housemates of his apartment house, Diallo is set on getting a breath of fresh air. He stands at the top of the steps to his own home, leaning against the wall and looking up and down the block. There is nothing noteworthy about this situation until an unmarked police vehicle turns up on Wheeler Avenue. One of the four officers in the Ford Taurus spots Diallo. At this time of the night, the officer assumes, that black man must be on the lookout for a burglar accomplice, who in the meantime is breaking into an apartment. And does he not fit the description of that serial rapist who made this neighborhood unsafe last year? The moment the car approaches, Diallo appears to step backward into the vestibule, like he does not want to be seen.

Context and environmental factors are coming into play. A black man who hangs out in a vestibule after midnight in one of the rougher neighborhoods of the Bronx is suspect for any white police officer. A cigarette could have made a difference. It might even have generated sympathy. In that case Diallo would have been held for one of those unnumerable men that are not allowed to smoke or vape indoors and have to take it outside. Subliminally or unconsciously, we perceive anything that is negative or regarded as negative much faster. Because of the lightning-fast process of selective information reduction, a young black man who is just getting some fresh air ends up in the category of suspicious individuals. This assumption is reinforced into that of a criminal when he accidentally steps back just as the car passes by. Because of the availability heuristic and confirmation bias, the police officer can only think that that guy must be up to something bad. And as for the resemblance to the rapist, *own-race bias* makes members of other racial or ethnic groups all look alike for persons that are not part of that group. On that spot, the police officer would have recognized the rapist in every black man of that stature, surely in the middle of the night, because perception is highly driven by context and expectations. When we lack social contacts with members of other racial groups, our brains do not get the opportunity to train themselves in recognizing their facial features. Just as Mike May could not recognize faces after his years of blindness, we are equally bad at distinguishing faces of persons belonging to other racial or ethnic groups.

The unmarked vehicle with the four police officers on board stops abruptly. They have already passed the flat, so they back up the car until they are in front of the vestibule where Diallo is standing. Two men get out of the car, but, wearing jeans, sweatshirts, and baseball caps, they look nothing like police officers. The four police officers are young, very young. The driver is 26, the colleague next to him is 35, and the two others in the back seat are 27. They also do not have many years of service on the force. They are new to the Bronx and to the Street Crime Unit. Their bulletproof vests inflate the chests of the two tall men that get out of the car. Not so long ago, someone Diallo knew had been robbed by a group of armed men. Diallo probably must have thought something along the lines of "Here comes trouble!" Intuition is always a combination of selective information reduction and unconscious or implicit associations. When worried about a situation, we always assume the worst possible outcome. What would you think if you were getting some fresh air in a neighborhood where the drug trade is thriving when suddenly a weekday Ford pulls over and two towering men get out and stride toward you? Something unexpected is happening, over which you have no control whatsoever. You are no longer at ease but in distress. You become anxious and your heart rate speeds up. There might be (mortal) danger here. The level of cortisol in your blood surges. Your brain has the choice between *fight*, *flee*, or *freeze*.

The two men address Diallo while holding up their badge: "Police. Can we have a word?" Their question remains unanswered. Later it emerged that Diallo had a stutter, the contribution of a genetic vulnerability. At that time, the young man is probably already so terrified that he no longer can utter a single word. He can no longer speak, but he can still flee. Diallo runs into the vestibule. The police officers rush up the four-step stoop and chase him down the darkened hallway. The heart rate and stress levels, which were already higher than normal in both Diallo and the police officers, are now skyrocketing. As all parties involved progress deeper into the dark vestibule, the level of the stress hormone cortisol keeps on rising in their bloodstream. By now perceptual narrowing is inevitable. Biochemically, their organisms are in a state of extreme activation. They suffer from *mind-blindness*. The outside door is closed. There Diallo gets stuck. With his body pressed against the door, he holds the doorknob in his right hand. He twists his body sideways and begins to dig at something in his pocket with his left hand. "Get your hands out of your pockets! Don't make me fucking kill you!" yells one of the police officers with his service weapon drawn. The tension becomes excruciating.

An expectation that is too strong can evoke a perception. Diallo starts removing a black object from his right pocket with his left hand. Based on his training and experience, the police officer can interpret this behavior and the tip of the black object that he notices in only one manner: the suspect is pulling his gun. The police officer yells out: "Gun! He has got a gun!" In any incident report of fatal arrests, the wallet of the victim looks invariably like a gun. Diallo now begins to raise the black object in their direction. One officer opens fire. At that moment, his colleague instinctively jumps backward off the step, firing as he falls. He lands on his back on the pavement. The officer in the vestibule assumes the suspect has shot his colleague. He keeps on shooting but now aims to hit the young black man. Seeing something or thinking something unconsciously is almost always at the origin of our behavior. In the meantime police officers number three and four have jumped out of the car and run toward the building. They see their colleagues shooting and, without questioning, do what their colleagues do: they also start to shoot. Group conformity. Diallo, meanwhile, is crouched beside the door and has his hand out. With his knees bent and back straight up, he looks exactly like someone who has made himself into a smaller target, just as police officer number three was taught during the countless target practices at the police academy. And in Diallo's hand, he also sees . . . a gun. At the trial he testified that at that very moment he was fearing for his life. You know all too well how straight that police officer still could think with all the cortisol in his brain.

From four semiautomatic handguns, 41 shots were fired in 2.5 seconds. Seventeen bullets hit the young black man. The entire incident, including the shooting, lasted only 7 seconds. As with the soldier of the Islamic Revolutionary Guard Corps who decided to fire two surface-to-air missiles on Wednesday, January 8, 2020, the time pressure on the police officers was significant. Reacting in a split second (or even a millisecond) is acting instinctively and thinking intuitively. As it would in Tehran 21 years later, in the following days and weeks in the Bronx, it became painfully clear that things were not what they had seemed. Diallo did not carry a gun, was no rapist, and stood not on the lookout for a burglar. He was just a young man of 23 with darker skin who went to get a breath of fresh air before going to sleep. By an unfortunate turn of events it became his last time.

Friday, January 24, 2020

I am finalizing my text for the Federation of European Fire Officers (FEU) and forward it for approval to the fire chief and the commander of my sector. Three days earlier, I found among my emails an appeal from the chairman of the FEU. They are working on a European fire safety action plan, and to support it they are gathering stories that describe the impact that fires have had on the lives of civilians and firefighters. I know such a story and immediately start to write. The deadline is tight. The text in English has to be submitted together with a photograph before the end of the month. For me, the imposed restriction in length is the hardest part. The testimony may not exceed 200 words. I make the most out of it and finish at exactly 200 words. After the weekend I send the text and the accompanying photograph to the chairman of the FEU:

> In the early hours of a Sunday in August 2019, the Fire and Rescue Service of South-West Limburg (Belgium) sets off for a fire in an abandoned supermarket. A fire as we have put out by the dozen. Why should it be different this time? Several teams are doing what they are trained for. The fire seemed under control when disaster struck. A blaze breaks through the ceiling and in no time the supermarket is completely engulfed in flames. Two colleagues are missing. They are stuck inside the building, but a sea of fire prevents us from entering. Never did we feel so helpless. Only hours later we managed to recover their lifeless bodies.
>
> On the day of their funeral, firefighters from all over the country bring in total silence an appropriate tribute as the ladder trucks carrying their coffins slowly pass by. Together with their loved ones, we give our colleagues all the recognition they deserve in a touching ceremony held at the fire station where everything began for them that ill-fated Sunday.
>
> And now, each time we set off in our fire truck, we think: "That was just not supposed to happen, not to our guys, not so close to home."

The following morning the chairman of the FEU thanks me for the testimonial. The feelings I evoke are not unknown to him. On his first day as a fire chief, disaster struck in his fire station, for at a fireground an experienced firefighter lost his life. As a fellow sufferer, he is willing to discuss that fatal incident and tell me how in due time everything fell more or less back into its rightful place. I seize the opportunity and we meet on March 6. On that day, I also inquire how the action plan is going. Unfortunately, the appeal yielded only three testimonials. That part of the plan was dropped.

EPILOGUE

> *Plato said that we are trapped inside a cave and know the world only through the shadows it casts on the wall. The skull is our cave, and mental representations are the shadows.*
>
> —Steven Pinker

Mens sana in corpore sano, or "a healthy mind in a healthy body." All along my study of classical humanities I was told this time and time again. However, in my younger days I did not fully grasp what it actually meant. Occasionally I have had an arm in plaster or undergone some minor surgery here or a major one there. Accidents happen. But an unhealthy mind, I had no idea what that would entail, let alone that one day I would fall victim to it. On Sunday night, August 11, 2019, that changed abruptly at 3:08 a.m.

Wednesday, September 11, 2019

We are exactly 1 month after the fire. A colleague mails me his account of that terrible night. A few days earlier he told me his story during a car ride to a team building just across the language border. He managed to fill in and enrich the gaps in my narrative with so much detail that I had no choice but to ask him to entrust his personal experience to paper and send it to me. I reply at once:

> Thanks for your story. I read it immediately. It was like reliving in fast forward the drive to last week's team building. All the details come back exactly as you told them last Friday.
>
> As for my story, it is already 44 pages long and most of it deals with my emotional state the days following the fire. There even was a day I thought I was going mad. Sooner or later I have to summarize it, and filter out the narrative of the fire. The day it is ready, I will send it to you.

Now, more than 4 years later, when this book is finally to be published, the above email almost seems prophetic. The story of the fire is chapter 1 of this work. Apparently my ambition ended there in early September 2019. At that time, I could have never imagined that one day I would add to that narrative 67 pages on the process of decision-making at a fireground. I was even less aware that thinking about thinking would pull me even more pages deep into the inner workings of the unconscious mind.

POSTSCRIPT

Wednesday, May 11, 2022

Again the fatal fire of Beringen makes news headlines. Yesterday a breakthrough was achieved in the judicial inquiry. Four suspects have been arrested.

At the time of the fire, three of them were minors. The fourth suspect had just turned 18. That night in 2019, three of them were inside the building. It was their regular hangout spot to chill out among friends, smoke a cigarette, or pass time. And if it got cold, they lit a fire.

In December 2018 they set fire to a bunch of papers in a plastic bowl. But that bowl started to melt, and in a hurry they tried to stamp out the fire.

Outside the building a local resident spotted the smoke and notified the fire service.

In August 2019 they were back in the building, which they knew like the back of their hands. This time they got bored and began to play with fire, literally. When they left the building later that night, unfortunately not all the fire had been extinguished properly.

ACKNOWLEDGMENTS

Writing a book is like building a house: you cannot do it all by yourself. Without the continual support of many, I would never have been able to even start writing on the fire of Beringen. Everyone who has supported me, or tried to do that, when I was going through rough times, I hereby thank you sincerely. Several persons went a step further and contributed directly or indirectly to this work. The risk with acknowledgments is forgetting someone. My sincere apologies should I have overlooked you.

Without my colleague Walter Cuypers of the Fire and Rescue Service of North-Limburg and his wife Ingrid Bergmans, this book would indisputably have another look and feel. Already during that terrible night, Walter began to get me through my darkest hours, and he continued to do so for days, weeks, even months until I was more or less back on track. And subsequently he transformed himself into my literary agent. Only after his approval did I dare to send the various chapters to colleagues for feedback. Behind every great man stands a strong woman. This is also the case at the Cuypers' home in Lommel. It soon became clear that most of the proposed corrections were not from the man of the house, but from his wife. If there is one thing fire officers are good at, then it is for sure delegating work. Wordings and ideas still plagued by all possible kinds of errors were pointed out with a merciless precision by Ingrid. This book would not have sounded half as good without her.

The first positive reaction to an embryonic part of this work also came from the Fire and Rescue Service of North-Limburg. The enthusiastic email of An Bollen told me that I was on the right track and had to keep on digging. Colleagues Arno Barzan, Marc Ceyssens, Peter Huysmans, Luc Raymaekers, Hugo Simons, Martijn Stassen, Ken Vaes, and Werner Van Kerckhove were my sounding board within the Fire and Rescue Service of South-West Limburg. Ideas can be sharpened by email exchange or during training, but a far more enjoyable manner is over a fine meal in a nice restaurant. I have expertly ruined many lunches with my endless babbling about the impact of unconscious thought rocesses on incident commanding at a fireground. I look forward to making it up with another lunch, but this time I am not going to say a word about the brain and the fascinating working of neurotransmitters. Talk about a challenge!

Several colleagues from the fire services of Aarschot, Beringen, Tessenderlo, the Fire and Rescue Service of North-Limburg, and the Fire and Rescue Service of East-Limburg literally had to sweat out the parts that in the end crystallized into this book. Some students of PLOT, the fire academy of the province of Limburg, were fellow sufferers. Their comment that parts of my class were ethereal made abundantly clear that a lot of fine-tuning was still awaiting me.

I offer my sincerest appreciation to my colleagues Marc Celis and Bert Gernaey, who have provided me their accounts of that ill-fated night. Thanks to their clear view on the matter, I was able to get some details sharper than several of my own blurred impressions.

Psychologist Marieke Impens lifted a tip from the veil that hangs over the working of the human mind. It led to more than either of us could have imagined. The egg timer model that I use at the end of chapter 2 to illustrate the recovery from an acute stress disorder comes from her. Better well stolen than badly invented!

Within the Network Fire Service (*Netwerk brandweer*) of the Association for Flemish Cities and Municipalities (*Vereniging voor Vlaamse Steden en Gemeenten*), I am indebted to Toon Dumoulin, Karel Lambert, and Bert Brugghemans. Toon digitized the plan of the fireground. An addition to the current model of reading the fire is better checked first with Karel, the subject matter expert. When I circulated my first draft of this book, Bert intuitively sensed that a part on the unconscious mind had to be added. His suggestion is at the origin of chapter 3. He probably would have never made that suggestion should he have known that the common thread through that chapter would be history.

I thank Joke Steyvers of the fire academy PLOT not only for the coffee, which unfortunately was spilled over a draft version of chapter 2, but also for the interesting intellectual ping-pong on white and black swans we have played in the lobby of the PXL in Hasselt. When you have tried to teach college students the basics of Karl Popper's thinking in your previous life, then you are always one step ahead with me.

Professor Dirk Hermans pointed out to me that the relationship between the squirts of cortisol in our blood and the learning capacity is curvilinear. He suggested it—how could it be otherwise—during a phone call. The days following the fire, I was constantly making phone calls. If they were not about how the mind works, then I was phoning the former promoter of my PhD to exchange ideas on the lives of saints in the early Christian Near East and the Middle Ages. A special acknowledgment to Professor Willy Clarysse for having provided additional information that allowed me to somewhat substantiate my wild idea on posttraumatic growth in the Middle Ages.

Journalist Hans Put of *Internetgazet Beringen* gave me access to his rich photo archive. The striking images on the cover testify to a trained photographic eye.

Ludo Laenen, Marcel Nickmans, and Marcel's wife Greta Bruyninx form a category on their own. Ludo kept on running the newspaper shop on the corner of Coal Mine Lane and Saint Barbara Street till his sudden and totally unexpected death in March 2022. With the ease and calm that typified him, he answered my numerous questions on the building adjacent to his newspaper shop and the various fires that over the course of time had wrecked it completely. In their turn, Marcel and Greta cherish fond memories of the supermarket Nopri and later Delhaize, which they had run for years on Coal Mine Lane. We met at the silent walk of August 18, 2019. In my search for information on the building's past, I paid them a visit in Stevoort. At their kitchen table I obtained firsthand information that I could not have found elsewhere. Ludo, Marcel, and Greta were so kind that I was delighted to receive all of them together with their families on Sunday, March 8, 2020, in the fire station of Beringen to talk about the fire and guide their grandchildren round the fire trucks in the parking lot. I will never forget that Sunday afternoon, not only for the masses of Easter eggs and homemade cookies they brought along but also for what happened next in Belgium. One weekend later, we all sat at home as the country went into lockdown due to COVID-19.

And last but not least, I am extremely grateful to my wife, Grace, and our two sons, Jonathan and Matthias. In the late morning of Sunday, August 11, 2019, a different partner and father respectively came into their lives from the one they had said goodbye to earlier that weekend. In the days, weeks, and months after the fire, they had to face the whims and fancies of a housemate in the most confusing period of his life. Fortunately, our bond has only grown stronger through that ordeal. Even more than before that fateful night, my wife and our two sons bring color to my life. Even lots of color. And for that my thanks are boundless.

After a turbulent period, my mind is finally coming to peace. Phew, my search for answers seems to be over.

GLOSSARY

adaptive unconscious: Most things you do (decisions, choices, actions), you do effortlessly and automatically. You are affected by everything that surrounds you, and you adjust to circumstances unconsciously.

apophenia: Inferring causal links or perceiving patterns and sequences where there are in fact only random processes.

availability bias or availability heuristic: How easily something comes to mind is based on how frequently you have seen or heard it in the past and how likely you think it is that it will happen again.

BE-SAHF model: The letters stand for building, environment, smoke, air, heat, and flame. On the basis of this empirical model, a fireground commander can form an impression of the development phase the fire is in and predict its future course.

bias: Prejudice or fallacy that causes you to process only the information that suits your way of thinking. See also availability bias, confirmation bias, egocentric bias, hindsight bias, and own-race bias.

bollandists: The Bollandist Society is an institution, founded by the Jesuits, dedicated to the critical study of Christian saints and their lives.

bottom-up processing: The sensory input from the external environment and the sense of the internal physiological condition of the body that enters various brain regions for further analysis and processing.

cockloft: Term used in the U.S. for the enclosed space between the finished ceiling and the underside of a flat roof.

cocktail party effect: Name for the phenomenon whereby in a crowded and noisy environment you can focus on the voice of a particular person and filter out everything else.

cognitive therapy: Therapy based on the idea that the interpretation of an event, rather than the event itself, determines your peace of mind. Negatively oriented interpretations lead to bad moods and incorrect or useless interpretations.

confirmation bias: The innate tendency to look only for those cues that confirm your ideas, strategy, or mental model and disregard the ones that do not match.

continuity hypothesis: The supposition that there is a fixed line in the behavior of vertebrates on evolutionary grounds.

cortisol: Also known as the stress hormone. Hormone produced in the adrenal glands in response to anxiety or stress.

curvilinear relationship: Relationship in which one variable is related to another variable, but in a graph the curve has the shape of an inverted U rather than a straight line.

dispositional or perceptual representations: Potential patterns of neural activity in small groups of brain cells. This operating principle is similar to the settings of a digital camera. Instead of just a few hundred settings, the brain has tens of billions of such settings that interact with one another.

dopamine: Neurotransmitter or molecule used by the nervous system to transmit messages. In the mesolimbic reward circuit of the brain this chemical messenger is released in anticipation of pleasure when unexpected possibilities arise.

egocentric bias or ego bias of memory: Prejudice or fallacy that makes your part in events appear more important than actually was the case. Hindsight bias and own-race bias fall under this heading.

empirical model: Scientific research method in which direct, personal observation is used to formulate a hypothesis, after which the formulated theory is tested in experiments to verify or refute it.

falsification: The possibility to find proof to the contrary of a particular theory. A single case can disconfirm a generally accepted theory derived from countless confirmatory examples.

fire gas ignition: A fire development in which smoke fumes that have accumulated in a space near the fire ignite (e.g., suspended ceiling or ducts).

frontal cerebral cortex: The cortex is the massively folded outer layer that surrounds the cerebral hemispheres, where information from the senses or inside the body is received, analyzed, and interpreted and then converted into predictions, thoughts, feelings, and behavior. The cortex of the frontal lobe allows you to solve problems rationally, think ahead, plan for the future, and repress impulses.

Geschwind syndrome: A group of behavioral symptoms that occur in individuals suffering from temporal lobe epilepsy. This rare mental health condition can profoundly alter one's personality. For example, individuals can become hyperreligious, develop an atypical sexuality, or experience a compulsive urge to write.

hindsight bias: The natural tendency to overestimate one's ability to predict events after the fact. This innate bias or fallacy falls under the heading of egocentric bias.

hineininterpretierung: Any form of "reading what is not there or seeing what is not there," because people tend to get stuck in their own mental model or framework. For example, we explain historical events based on the outcome that is now known.

inattentional blindness or illusion of attention: The phenomenon whereby you are so focused on one aspect of your current environment that you overlook other objects, even when they are salient. You are effectively blind to things you do not expect to see and you do not devote attention to.

intrusive symptoms: This category consists not only of thoughts and memories but also of objects, sounds, and images that are reminiscent of a traumatic event. The symptoms can manifest for 1 or more months after the traumatic experience. Occasionally they may keep intruding for one's entire life.

limbic system: A collection of brain areas located at the edge of the forebrain (limbus is Latin for "edge, border"). This area is involved in, among other things, emotions and memory. The amygdala, hippocampus, and thalamus are parts of this brain region.

mind-blindness: A kind of context blindness. Extreme stress and anxiety distort the functioning of your conscious mental processes to such an extent that you fail to see the whole picture, focusing only on the details that frighten you. All context is lost and perceptual narrowing sets in.

narrative fallacy: The irresistible tendency we have to fit a story or pattern to a series of connected or disconnected events.

neurochemical algorithms: An algorithm is a process or a set of rules to be followed when performing calculations or other problem-solving operations. Neurochemistry is the chemical processes in our nervous system.

neuron: A nerve cell that is specialized in receiving, processing, and transmitting information. These cells are found not only in the brain but also in the rest of the human body. The number of neurons in our brain is on the order of 86 billion. They are interconnected, and together they form the autonomic nervous system that enables you to perform various functions.

neurotransmitter: A molecule or chemical used by the nervous system to transmit messages and control bodily functions. These molecules are the brain's chemical messengers.

noradrenaline: A hormone and neurotransmitter that is released during stress and anxiety. It seriously distorts our thought processes.

own-race bias: Own-race faces are better recognized than faces of another race. This bias or natural tendency falls under the heading of egocentric bias.

pluralistic ignorance or advantage of conformity: The tendency to conform to the group norm and to do what everybody does, whether or not you agree to it.

posttraumatic growth: The process that makes you emerge from a traumatic event stronger after you have managed to come to terms with what has happened and direct your life away from it.

posttraumatic stress disorder (PTSD): A mental condition that can arise after having experienced a shocking event. The psychological consequences of such a trauma can be long-lasting.

priming: The continuous but subtle influencing of the adaptive unconscious by all things and persons that surround you in everyday life.

retrospective distortion: Each time you retrieve a memory, you modify the course of the event. In retrospect an event always unfolds more clearly and orderly than it actually did.

selective information reduction: The ability of the adaptive unconscious to rapidly discover or recognize cues in situations or behavior by selectively reducing the perceived reality to thin slices of information.

situational awareness: The ability to perceive what is happening around you and understand it in relation to how time progresses.

standard operating procedure (SOP): A procedure that experts in firefighting techniques have developed to handle a particular kind of incident and that is practiced repeatedly by firefighters.

stay-put policy: The point of departure from the Standard Operating Procedure of the London Fire Brigade for fires in high-rise residential blocks until the Grenfell Tower Fire (June 2017). Residents were asked to stay in their flat with the door shut and wait until the fire service had done their job.

striatum: A brain region in the forebrain that is essential for movement, among other things, and wherein the neurotransmitter dopamine plays an important role.

System 1 or automatic mental processes: Your reflexes, intuitions and automatic thinking patterns. They operate quickly and automatically, with little or no effort, and no sense of voluntary control.

- A **reflex** is an innate behavioral pattern that has its origin in an automatic physiological reaction you got from Mother Nature at birth. For example, the Fs of fight-flight-freeze were inscribed in the human brain during our long evolutionary history, and thus this behavior is now passed down to the next generation through our DNA.
- An **intuition** is a learned thinking pattern with which you quickly interpret everything that happens around you without thinking too hard.

- **Automatic thinking patterns** are skills that have been taught but are now automatic because they have been mastered through frequent use and experience (e.g., reading or driving).

System 2 or conscious mental processes: All your effortful mental activities that require attention. They cost (mental) energy and are slow. These include, for example, complex computations, planning, self-control, and decision-making.

top-down processing: All the information that different brain regions add to sensory input and/or information from inside the body based on prior knowledge, past experience, and expectations during processing.

WYSIATI: "What You See Is All There Is." Acronym coined by Kahneman for the phenomenon whereby you instantaneously and intuitively draw conclusions based on limited evidence and partial information without questioning. Variants on this theme are "What you see is what you do" (J. Bargh) and "What you see is what you get" (S. Sommers).

NOTES

INTRODUCTION

1 Merlevede, *Vitamines en hormonen*, 235.

CHAPTER 2

1 Taleb, *Black Swan*, 115–16.

2 Harari, *Sapiens*, 258.

3 rayOman, "Charlie Chaplin Optic Illusion," May 16, 2006, https://youtu.be/QbKw0_v2clo.

4 Transport for London, "Test Your Awareness: Do the Test," public safety advertisement, https://youtu.be/Ahg6qcgoay4. See also Chabris and Simons, *Invisible Gorilla*, 5–11.

5 Kahneman, *Thinking, Fast and Slow*, 275.

6 Aleman, *Hersenspinsels*, text on the back cover.

7 Kahneman, *Thinking, Fast and Slow,* 44–45.

8 Cohen-Hatton, Butler, and Honey, "Investigation of Operational Decision Making," 802–03.

9 Similar airplane crashes are analyzed in Klein, *Sources of Power*, 52–54, 79–91, and Klein, *Streetlights and Shadows*, 264–66. See also Pinker, *Rationality*, 201–26 (signal detection and statistical decision theory).

10 Pavel Aksenov, "Stanislav Petrov: The Man Who May Have Saved the World," *BBC News,* September 26, 2013, https://www.bbc.com/news/world-europe-24280831.

11 Bargh, *Before You Know It*, 142.

12 Kahneman, *Thinking, Fast and Slow*, 416.

13 Grenfell Tower Inquiry, *Grenfell Tower Inquiry*, 14.

14 Van der Zee, *Het vitale brein*, 142–57.

15 Dijksterhuis, *Het slimme onbewuste*, 139–59.

16 Damasio, *Descartes' Error*, 27–28.

17 Nesse, *Good Reasons for Bad Feelings*, 74.

18 American Psychiatric Association, *Alles over psychische stoornissen*, 158–62.

19 Kahneman, Sibony, and Sunstein, *Noise*, 31.

20 NIOSH, *Two Career Fire Fighters Die,* 1.

21 Gryspeert, Lambert, and Merci, *Onderzoek en evaluatie*, 29.

22 Tremblay and Chiasson, *Accident mortel*, 24.

23 Lieberman and Long, *Molecule of More*, 125.

CHAPTER 3

1 Descartes, *Discours de la méthode*, 23.

2 Descartes, *Objections to the Meditations*, 75.

3 Dijksterhuis, *Het slimme onbewuste*, 195 and 203.

4 According to Aleman, it is simply easier to take the first or last pair of the row. Interestingly enough, first pair on the far left was chosen the least often. I do not find his better explanations convincing. Aleman, *Je brein de baas*, 23–24.

5 Bargh, *Before You Know It*, 129.

6 Godfrey-Smith, *Metazoa*, 20.

7 Dijksterhuis, *Het slimme onbewuste*, 215.

8 Gladwell, *Blink*, 53–58.

9 Dijksterhuis, *Het slimme onbewuste*, 116.

10 Vlaamse Statistische Autoriteit, "Statistiek Vlaanderen," November 17, 2022, https://www.vlaanderen.be/statistiek-vlaanderen/bevolking/echtscheidingen.

11 The unconscious thought theory is not undisputed; see the criticism in Aleman, *Je brein de baas*, 28–31.

12 Magiels, *Al te gek*, 100.

13 Magiels, *Al te gek*, 100.

14 American Psychiatric Association, *Alles over psychische stoornissen*, 145–47.

15 Knapp, "Post-Traumatic Growth," 39–48.

16 Harari, *Homo Deus*, 293.

17 *Ignatius*, 12.

18 All information relating to the tragic death of Diallo is taken from Gladwell, *Blink,* 90–94, 115–16.

19 See, e.g., Gladwell, *Blink,* 37–39; Dijksterhuis, *Het slimme onbewuste*, 97–100; and particularly Sommers, *Situations Matter*, 250–67. According to Aleman, *Je brein de baas*, 26–27, a meta-analysis on the Implicit Association Test has shown that there is no connection between unconscious racial attitudes and discriminatory behavior. With your conscious awareness (System 2) you can override the unconscious thought processes of System 1. That is not the problem. The problem is that your conscious mind hardly checks those fast unconscious processes or stereotypes (WYSIATI), and therefore they still direct your behavior. A similar analysis is found in Bargh, *Before You Know It*, pp. 69–71, 88–93.

SELECTED BIBLIOGRAPHY

THE WORKING OF THE BRAIN

Aamodt, S., and S. Wang. *Welcome to Your Brain: Why You Lose Your Car Keys but Never Forget How to Drive and Other Puzzles of Everyday Behavior.* London: Rider, 2008.

Aleman, A. *Hersenspinsels: Waarom we dingen zien, horen en denken die er niet zijn.* Amsterdam-Antwerp: Atlas, 2011.

———. *Het seniorenbrein: De ontwikkeling van onze hersenen na ons vijftigste.* 17th ed. Amsterdam: Olympus, 2019.

———. *Je brein de baas: Over de rol van bewust denken.* Amsterdam-Antwerp: Olympus, 2017.

Ashwell, K. *The Brain Book: Development, Function, Disorder, Health.* 2nd ed. Richmond Hill, ON: Firefly Books, 2019.

Barret, L. *How Emotions Are Made: The Secret Life of the Brain.* London: Pan, 2018.

Bryden, D. W., J. I. Tilghman, and S. R. Hinds, II. "Blast-Related Traumatic Brain Injury: Current Concepts and Research Considerations." *Journal of Experimental Neuroscience*, no. 13 (2019): 1–11. https://doi.org/10.1177/1179069519872213.

Burnett, D. *The Happy Brain: The Science of Where Happiness Comes From, and Why.* London: Faber and Faber, 2018.

———. *The Idiot Brain: A Neuroscientist Explains What Your Head Is Really Up To*. London: Faber and Faber, 2017.

Cobb, M. *The Idea of the Brain: A History*. London: Profile Books, 2021.

Damasio, A. *Descartes' Error: Emotion, Reason, and the Human Brain*. New York: Harper, 1995.

Dijksterhuis, A. *Het slimme onbewuste: Denken met gevoel*. Amsterdam: Bert Bakker, 2007.

Eagleman, D. *The Brain: The Story of You*. Edinburgh: Canongate Books, 2016.

Haas, F. H. "Phineas Gage and the Science of Brain Localisation." *Journal of Neurology, Neurosurgery & Psychiatry* 71, no. 6 (2001): 761.

Lieberman, D. Z., and M. E. Long. *The Molecule of More: How a Single Chemical in Your Brain Drives Love, Sex and Creativity—and Will Determine the Fate of the Human Race*. Dallas: BenBella Books, 2019.

Pennartz, C. M. A. *The Brain's Representational Power: On Consciousness and the Integration of Modalities*. Massachusetts: MIT Press, 2015.

Sacks, O. *The Man Who Mistook His Wife for a Hat*. London: Picador, 1986.

Seth, A. *Being You: A New Science of Consciousness*. London: Faber and Faber, 2021.

Sommer, I. *Haperende hersenen*. 9th ed. Amsterdam: Balans, 2018.

Swaab, D. *Ons creatieve brein: Hoe mens en wereld elkaar maken*. Amsterdam-Antwerp: Atlas, 2016.

———. *Wij zijn ons brein: Van baarmoeder tot alzheimer*. 10th ed. Amsterdam-Antwerp: Atlas, 2011.

Talsma, D., and J. Taquet. *De glazen bol in ons hoofd: Hoe ons brein het verleden gebruikt om de toekomst te voorspellen*. Ghent: Owl Press, 2021.

van der Zee, E. *Het vitale brein: Hoe de hersenen sterk en dynamisch te houden*. Amsterdam: Bakker, 2012.

THINKING: PHILOSOPHY AND PSYCHOLOGY

American Psychiatric Association. *Alles over psychische stoornissen*. Amsterdam: Boom, 2016.

Baars, B. J. *In the Theater of Consciousness: The Workspace of the Mind*. Oxford: Oxford University Press, 1997.

Bargh, J. *Before You Know It: The Unconscious Reasons We Do What We Do.* London: William Heinemann, 2017.

van Bendegem, J. P. "L'homme entre magie et science." In *Magie blanche, magie noire: Catalogue de l'exposition à la Galerie CGER à Bruxelles du 20 janvier - 7 mai 1995*, edited by I. De Meûter and M. Poriau, 227–38. Brussels: CGER, 1995.

Brafman, O., and R. Brafman. *Sway: The Irresistible Pull of Irrational Behavior.* New York: Doubleday, 2008.

van den Broek, E., and T. Den Heijer. *Het bromvliegeffect: Alledaagse fenomenen die stiekem je gedrag sturen.* 4th ed. Amsterdam: Unieboek, 2021.

Chabris, C., and D. Simons. *The Invisible Gorilla: And Other Ways Our Intuitions Deceive Us.* New York: Crown, 2010.

Descartes, R. *Discours de la méthode.* Online ed. University of Quebec at Chicoutimi, 1637. http://classiques.uqac.ca/classiques/Descartes/discours_methode/discours_methode.html.

———. *Objections to the Meditations and Descartes's Replies.* Translated by J. Bennet. 1641. https://www.earlymoderntexts.com/assets/pdfs/descartes1642.pdf.

Drew, T., M. L. H. Vo, and J. M. Wolfe. "The Invisible Gorilla Strikes Again: Sustained Inattentional Blindness in Expert Observers." *Psychological Science* 24, no. 9 (2003): 1848–53.

Gawande, A. *The Checklist Manifesto: How to Get Things Right.* London: Profile Books, 2011.

Gladwell, M. *Blink: The Power of Thinking Without Thinking.* New York: Little, Brown, 2005.

Godfrey-Smith, P. *Metazoa: Animal Minds and the Birth of Consciousness.* London: HarperCollins, 2020.

Hoogendijk, W., and W. de Rek, *Van big bang tot burn-out: Het grote verhaal over stress.* Amsterdam: Balans, 2017.

Kahneman, D. *Thinking, Fast and Slow.* New York: Farrar, Straus, and Giroux, 2011.

Kahneman, D., and G. Klein. "Conditions for Intuitive Expertise: A Failure to Disagree." *American Psychologist* 64, no. 6 (2009): 515–26.

Kahneman, D., O. Sibony, and C. R. Sunstein. *Noise: A Flaw in Human Judgment.* London: William Collins, 2022.

Klein, G.A. "A Recognition-Primed Decision (RPD) Model of Rapid Decision Making." In *Decision Making in Action: Models and Methods*, edited by G.A. Klein, J. Orasanu, R. Calderwood, and C.E. Zsambok, 137–47. Norwood: Praeger Publishers, 1993.

———. *Sources of Power: How People Make Decisions.* 20th Anniversary Edition. Massachusetts: MIT Press, 2017.

———. *Streetlights and Shadows: Searching for the Keys to Adaptive Decision Making*. Cambridge and London: Bradford Books, 2011.

———. *The Power of Intuition: How to Use Gut Feelings to Make Better Decisions at Work*. New York: Crown Currency, 2004.

van der Kolk, B. A. *The Body Keeps the Score: Brain, Mind, and Body in the Healing of Trauma*. New York: Penguin, 2014.

Lauwers, P. *Geloof, ongeloof en bijgeloof: Een godsdienstpsychologisch onderzoek naar de voorsocratische theorieën over de oorsprong van de godsdienst*. Antwerp: Standaard Wetenschappelijke Uitgeverij, 1974.

Lehrer, J. *How We Decide*. New York: Houghton Mifflin Harcourt, 2009.

Magiels, G. *Al te gek: Psychose tussen brein en samenleving*. Tielt: Lanoo, 2016.

Merlevede, W. *Vitamines en hormonen*. Leuven: Leuven University Press, 1989.

Natterson-Horowitz, B., and K. Bowers. *Zoobiquity: What Animals Can Teach Us About Health and the Science of Healing*. New York: Alfred A. Knopf, 2012.

Nesse, R. M. *Good Reasons for Bad Feelings: Insights From the Frontier of Evolutionary Psychiatry*. New York: Penguin, 2019.

Orasanu, J., and E. Salas. "Team Decision Making in Complex Environments." In *Decision Making in Action: Models and Methods*, edited by G.A. Klein, J. Orasanu, R. Calderwood, and C.E. Zsambok, 327–45. Norwood: Praeger Publishers, 1993.

Pinker, S. *How the Mind Works*. New York: W. W. Norton, 2009.

———. *Rationality: What It Is, Why It Seems Scarce, Why It Matters*. Dublin: Allen Lane, 2021.

Pluchino, E., A. E. Biondo, and A. Rapisarda. "Talent versus Luck: The Role of Randomness in Success and Failure." *Advances in Complex Systems* 21, no. 3n04 (2018): 1850014.

Popper, K. *The Logic of Scientific Discovery*. London: Routledge, 2002.

Reason, J. *Human Error.* 20th ed. Cambridge: Cambridge University Press, 2009.

Rosenhan, D. L. "On Being Sane in Insane Places." *Science* 179, no. 4070 (1973): 250–58.

Sommers, S. *Situations Matter: Understanding How Context Transforms Your World*. New York: Riverhead Books, 2011.

de Spinoza, B. *Tractatus theologico-politicus*. Hamburg: Henricum Künraht, 1670.

Taleb, N. N. *The Black Swan: The Impact of the Highly Improbable*. 2nd ed. New York: Random House, 2010.

Turnbull, G. *Trauma: From Lockerbie to 7/7: How Trauma Affects Our Minds and How We Fight Back*. London: Bantam Press, 2011.

Wilson, T. D. *Strangers to Ourselves: Discovering the Adaptive Unconscious*. Cambridge, MA: Harvard University Press, 2002.

HISTORY, GENETICS AND BEHAVIORAL BIOLOGY

Benneyworth, G. C. "Land, Labour, War and Displacement: A History of Four Black Concentration Camps in the South African War (1899–1902)." *Historia* 64, no. 2 (2019): 1–20.

Boncquet, P., and G. Verschatse. *Het Laatste Avondmaal naar Leonardo da Vinci: Een wonderlijke geschiedenis*. Gorredijk: Sterck and de Vreese, 2019.

Bregman, R. *De meeste mensen deugen: Een nieuwe geschiedenis van de mens*. N.p., 2019.

Dawkins, R. *The Selfish Gene*. 2nd ed. Oxford: Oxford University Press, 2016.

Dawkins, R., and Y. Wong. *The Ancestor's Tale: A Pilgrimage to the Dawn of Life*. London: Orion, 2007.

Dessein, J. *In de kijker . . . norbertijnerabdijen*. Averbode: Altiora/Vakantie-genoegens, 1988.

van Duppen, D., and J. Hoebeke. *De supersamenwerker*. Berchem: EPO, 2016.

Erman, A., and H. Schäfer. "Der angebliche ägyptische Bericht über die Umschiffung Afrikas." *Sitzungsberichte der Königlich Preussischen Akademie der Wissenschaften*, no. 39 (1908): 956–67.

Gertzen, T. L. "Der angebliche Bericht über die Umschiffung Afrikas." *Zeitschrift der Ägyptische Sprache und Altertumskunde*, no. 137 (2010): 104–12.

Geybels, H. "Kalender van heiligen en tradities in Vlaanderen: Herfst & winter." In *Heiligen en tradities in Vlaanderen: Herfst & winter*, edited by H. Geybels, 8–93. Leuven-Antwerp: Davidsfonds, 2018.

Harari, Y. N. *Homo Deus: A Brief History of Tomorrow*. London: Vintage Books, 2017.

———. *Sapiens: A Brief History of Humankind*. London: Vintage Books, 2015.

Herodotus. *The Histories*. Rev. ed. Translated by A. de Sélincourt. London: Penguin Classics, 1996.

van der Horst, P. W. *De woestijnvaders: Levensverhalen van kluizenaars uit het vroege christendom*. Amsterdam: Prometheus, 1998.

Ignatius, zijn orde en zijn geest. Exhibition brochure. Leuven, University Library, 21.03-18.04.1991.

de Meulder, B. *De kampen van Kongo: Arbeid, kapitaal en rasveredeling in de koloniale planning*. Antwerp: Meulenhoff, 1996.

Nelissen, M. *De brein machine: De biologische wortels van emoties en gevoelens: Een darwinistische kijk*. Tielt: Lannoo, 2008.

Pakenham, T. *The Scramble for Africa: The White Man's Conquest of the Dark Continent from 1876 to 1912*. New York: HaperCollins, 1991.

Rutherford, A. *A Brief History of Everyone Who Ever Lived: The Stories in Our Genes*. London: Weidenfeld & Nicolson, 2016.

Sterckx, A. "Geneeskunde op basis van waarneming en ondervinding: Een uitvinding van de 16de eeuw?" In *Renaissance in meervoud*, edited by J. Janssens and C. Matheeusen, 248–52. Leuven: Davidsfonds 1995.

Suhr, D. *Das Mosaik der Menschwerdung: Vom aufrechten Gang zur Eroberung der Erde: Humanevolution im Überblick*. Berlin-Heidelberg: Springer, 2018.

Vangroenweghe, D. *Rood rubber: Leopold II en zijn Kongo*. Leuven: Van Halewyck, 2004.

Vanhemelrijck, F. *Marginalen in de geschiedenis: Over beulen, hoeren, zigeuners en andere zondebokken*. Leuven: Davidsfonds, 2004.

de Waal, F. *Are We Smart Enough to Know How Smart Animals Are?* London: Granta Books, 2016.

Waddell, H., trans. *The Desert Fathers*. New York: Vintage Books, 1998.

van de Walle, B. "L'histoire véridique des faux scarabées de Néchao." In *La collection égyptienne: Les étapes marquantes de son développement*, edited by B. van de Walle, L. Limme, and H. de Meulenaere, 81–92. Brussels: Musées royaux d'art et d'histoire, 1980.

FIRE (SERVICE) AND INCIDENTS

Borhez, S. "Etanchéité à l'air des habitations et pression induite en cas d'incendie: Cocktail dangereux pour les occupants!" *Fireforum Magazine*, no. 82 (2022): 42–47.

Brugghemans, B., and H. Marynissen. "The Process of Decision-Making in a Fast-Burning Crisis Situation: A Multiple Sequence Approach of Decisiveness." *International Fire Service Journal of Leadership and Management*, no. 7 (2013): 41–50.

Coenen, M. "Brein en brand: De impact van stress." *Fireforum Magazine*, no. 85 (2022/2023): 36–43.

———. "Brein en brand: Denken onder druk." *De brandweer [m/v]*, no. 582 (2022): 21–27.

———. "Le cerveau et les incendies: L'instinct contre l'intuition." *Fire & Security Alert Magazine*, no. 28 (2022): 51–54, and no. 29 (2022): 16–20.

———. "Théorie relative à la prise de décisions par les pompiers." *Fireforum Magazine*, no. 80 (2021/2022): 7–15.

———. "Une théorie scientifique pour les pompiers: Le modèle 'I BE-SAHF.'" *Fireforum Magazine*, no. 75 (2020/2021): 8–18.

———. *Wat heb ik fout gedaan? Waarom een brandweerinterventie zo dramatisch kon eindigen.* Tielt: Lannoo, 2021.

Cohen-Hatton, S. R. *The Heat of the Moment: A Firefighter's Stories of Life and Death Decisions*. London: Transworld, 2020.

Cohen-Hatton, S. R., P. C. Butler, and R. C. Honey. "An Investigation of Operational Decision Making In Situ: Incident Command in the U.K. Fire and Rescue Service." *Human Factors* 57, no. 5 (2015): 793–804.

Driessens, J. "Weer een gevelbrand? De brand in de Torre dei Moro-toren in Milaan." *Fire & Security Alert Magazine*, no. 26 (2022): 6–11.

Evens, S. *De brand in de Innovation: De geschiedenis van de ramp die België veranderde.* Antwerp: Witsand, 2017.

Gasaway, R. B. *Situational Awareness for Emergency Response.* Tulsa: PennWell/Fire Engineering, 2013.

Godeysen, H., and H. Uhl. *155 Kriminalfall Kaprun.* Vienna: edition a, 2014.

Gowlett, J. A. J. "The Discovery of Fire by Humans: A Long and Convoluted Process." *Philosophical Transactions of the Royal Society B: Biological Sciences* 371, no. 1696 (2016): 20150164.

Grenfell Tower Inquiry. *Grenfell Tower Inquiry: Phase 1 Report.* 4 vols. United Kingdom: Her Majesty's Stationery Office, 2019.

Gryspeert, C., K. Lambert, and B. Merci. *Onderzoek en evaluatie van het incident in Beringen van 11 augustus 2019.* Algemene Directie Civiele Veiligheid. January 15, 2020.

Helsloot, I., E. Oomes, and R. Weewer. *Eindrapport: Evaluatie van de brand in De Punt op 9 mei 2008.* The Hague: Boom Juridische Uitgevers, 2009.

Hudson, J. "Self-Care and Emotional Wellness in the Fire Service." *Fire Engineering* 173, no. 12 (2020): 43–52.

Kernick, G. *Catastrophe and Systemic Change: Learning from the Grenfell Tower Fire and Other Disasters.* London: London Publishing Partnership, 2021.

Knapp, J. "Post-Traumatic Growth: Normal Reactions to Abnormal Incidents." *Fire Engineering* 173, no. 6 (2019): 39–48.

Lambert, K., and S. Baaij. *Brandverloop: Technisch bekeken, tactisch toegepast.* 2nd ed. The Hague: SDU Uitgevers, 2018.

Maclean, N. *Young Men and Fire.* 25th anniversary edition. Chicago: The University of Chicago Press, 2017.

National Institute for Occupational Safety and Health (NIOSH). *Two Career Fire Fighters Die while Making Initial Attack on a Restaurant Fire–Massachusetts.* NIOSH Fatality Assessment and Control Evaluation Report No. F2007-32. November 9, 2009. https://www.cdc.gov/niosh/fire/pdfs/face200732.pdf.

Norman, J. *Fire Officer's Handbook of Tactics.* 5th ed. Tulsa: Fire Engineering Books and Videos, 2019.

Pyne, S. J. *Fire: Nature and Culture.* London: Reaktion Books, 2012.

Rapport van de Raad voor de luchtvaart. No. 1979-3. The Hague, 1979.

Roitsch, P. A., G. L. Babcock, and W. W. Edmunds. *Human Factors Report on the Tenerife Accident*. Washington, DC: Air Line Pilots Association Engineering and Air Safety, 1978.

Shapiro, L. H. "Fire(fighter)-Related Human Behavior: Understanding Cognitive Limitations on the Fireground." *Fire Engineering* 175, no. 2 (2022): 34–40.

Tremblay, S., and Chiasson, A. *Accident mortel survenu à un travailleur dans un bâtiment résidentiel situé au 12 495 Langelier de l'arrondissement Montréal-Nord*. Investigation Report No. RAP0509812. Direction régionale de Montréal 2. 2008.

Weick, K. E. "The Vulnerable System: An Analysis of the Tenerife Air Disaster." *Journal of Management* 16, no. 3 (1990): 571–93.

INDEX